To Rupert
From Mum & Dad

Christmas 1974

GUIDE TO THE STARS

by

Patrick Moore, O.B.E., D.SC., F.R.A.S.

LUTTERWORTH PRESS
GUILDFORD AND LONDON

This completely revised and new edition
first published 1974

ISBN 0 7188 1782 6

Printed in Great Britain by
Ebenezer Baylis and Son Ltd.
The Trinity Press, Worcester, and London

CONTENTS

LIST OF PLATES

(Plates I to XII between pages 112 and 113)

FOREWORD

EVERYBODY HAS SEEN the stars—but by no means everybody knows just what a star is. What I have tried to do, in this book, is to give a general account of stellar astronomy, with emphasis upon the observational aspect. There are no new theories or world-shattering pronouncements here, and so the book will be of absolutely no interest to the well-informed reader; but I hope that it may help to introduce some beginners to the wonders of the night sky.

The original edition was published in 1960, and so much has happened since then that the text has been completely re-written. Quasars, pulsars and black holes, for instance, were unknown when the book first came out. By 1980 it will doubtless need revising again; meanwhile, I have done my best.

PATRICK MOORE

Selsey, *December 1973*

ACKNOWLEDGEMENTS

I WOULD LIKE to offer my thanks to the many people who have helped with the preparation of this book. In particular, I should like to thank Lawrence Clarke, who was responsible for all the line drawings, and Iain Nicolson, Lecturer in Astronomy at the Hatfield Polytechnic Observatory, who read the proofs of this book and made some most valuable comments, though he can on no account be held responsible for any errors or omissions in the text!

P.M.

Chapter One

THE SUNS OF SPACE

THERE ARE FEW sights more beautiful than a starlit sky on a dark, clear night. There are stars of all kinds—bright and dim, twinkling and almost steady, white and coloured; they form patterns and groups, and it is not hard to understand why our remote ancestors were so intrigued by them.

In our modern world there are some people who seldom see the stars at all. From a great city, such as London or New York, the sky is so brilliant with the glare of artificial lights that the stars are drowned. Even some of those people who live in the country often fail to take real notice of the stars; like trees, fields and roads, they are 'there', and are simply accepted as part of the natural scene. On the other hand, popular interest in astronomy today is certainly greater than it has ever been, and there is a general realization that a star is something more significant than a mere twinkling point.

I have no doubt that this surge of interest in astronomy is due in part to the space-research programmes. As we all know, men have reached the Moon; rocket probes have landed upon the planets Venus and Mars, and automatic vehicles have been sent even further afield, out toward the giant planets. On the whole, 'popular astronomy' has always tended to concentrate upon our own part of the universe—which means, in effect, the Solar System: the system made up of the Sun, the nine planets, the Moon and various other bodies such as comets and meteors. Certainly there can be no hope of sending space-probes out to the stars, at least by using methods available to us at the moment. It is pointless to talk about 'conquering the universe'; it would be rather like the attitude of a man who looks out of his office window in London, decides that he is capable of crossing the street to visit a house opposite, and then claims that he is well on the way to exploring the whole Earth.

What we have to remember is that our Solar System is extremely small on the cosmical scale; in fact, the more we learn about the universe, the more unimportant the Earth seems to become. We have come a long way from the time when men seriously believed our world to be the centre of all things. Even the Sun, which seems so glorious to us, is an ordinary star, and astronomers are unkind enough to classify it as a yellow dwarf. We know of one star which is at least a million times more luminous than the Sun, though it is so remote that with the naked eye it cannot be seen at all.

What I propose to do, in the present book, is to give a general outline of modern stellar astronomy—the study of the stars themselves, rather than taking much notice of the Solar System. I shall say very little about the planets, and virtually nothing about the Moon. However, I must deal at some length with the Sun, for the excellent reason that the Sun is a typical star and is the only one close enough to be studied in detail. But before going any further I must, I think, give a very brief outline of the basic principles of astronomy. Most people will be familiar with them already, but there is no harm in making sure.

In astronomy, we have to deal with immense distances and vast spans of time. The Sun, for instance, is 93,000,000 miles* from the Earth, and by everyday standards this is a very long way—so far that nobody can really appreciate it. (Personally, I am quite unable to realize what is meant by even one million miles, to say nothing of 93 million.) Yet astronomically the Sun is almost on our doorstep, and one of our very closest neighbours in space. We know that the values given by astronomers are of the right order, and all we can do is to accept them at their face value.

The Sun—a typical star, as I have said—is a globe of incandescent gas, with a diameter of 865,000 miles and a volume more than a million times greater than that of the Earth. Even its surface is extremely hot, at a temperature of 6,000 degrees Centigrade, and near the core its temperature rises to the fantastic value of about 14 million degrees C. Obviously, the Sun is self-luminous; it sends us virtually all our

* Or, if you like, 149,600,000 kilometres. At the moment people in general still think in inches, feet, yards and miles, despite all official efforts to 'go Metric' and lose our individuality. The change may well come; for the moment it seems best to keep to the familiar units. I have given temperatures in Centigrade rather than Fahrenheit, because the values are quite beyond our everyday experience in any case; and in official science, of course, Centigrade is always used.

light and heat, and without it we could not survive for a moment. Without the Sun, the Earth itself would not have come into being.

Round the Sun there move nine planets. These planets are divided into two well-marked groups. The inner part of the system contains four relatively small worlds, Mercury, Venus, the Earth and Mars; then comes a wide gap, in which move thousands of dwarf planets known as asteroids; and then we come to the giant planets Jupiter, Saturn, Uranus and Neptune. The ninth member of the system, Pluto, is included in the outer group, but is relatively small, and seems to be in a class of its own. The planets are not self-luminous; they shine because they reflect the light of the Sun—and in the fortunately improbable event of the Sun being snuffed out, the planets would promptly disappear from view. So too would the Moon, which is the Earth's companion in space, and keeps together with us as we journey round the Sun. Officially, the Moon is ranked as a satellite or secondary body accompanying the Earth. I have a suspicion that the Earth-Moon system should be better regarded as a double planet, but to discuss the problem here would be too much of a digression. Neither will I do more than mention the satellites of the other planets, some of which are larger than our Moon. Jupiter, the senior planet of the Solar System, has as many as twelve attendants, though admittedly eight of them are very small.

The diameter of the Earth is just under 8,000 miles. That of Jupiter is well over 80,000 miles. Large though Jupiter is in comparison with the Earth, it is still tiny when compared with the Sun, and there is a fundamental difference between a planet and a star. The ancient astronomers knew this quite well, if only because the planets were seen to behave in a most unstarlike manner. They move about among the star-groups or constellations, whereas the stars themselves remain fixed with respect to each other. Even today we still sometimes meet with the old, rather misleading term of Fixed Stars.

I must say more about this, because it is of such vital importance. Though the stars are not really fixed, and are moving through space in all sorts of directions at all sorts of speeds, they seem to keep to the same patterns over immensely long periods. The constellations which you can see tonight (clouds permitting!) are virtually the same as those which must have been seen by William the Conqueror, Julius Cæsar, Homer and the men of the Ice Age. To be accurate, differences would be found between the Ice Age sky and our own (remember, the Ice Age ended some 10,000 years ago), but the main patterns would still be recognizable, and there are no noticeable changes in the constellations over periods amounting to many lifetimes. This is simply because the stars are so far away.

To give a homely analogy, consider two bodies very near home: a sparrow flying about among the tree-tops, and a jet aircraft thousands of feet above the ground. Of the two, the sparrow will seem to be travelling much the quicker; the jet will crawl against the clouds. Yet in reality the jet is the faster of the two. No sparrow has much hope of breaking the sound barrier, and its movement is the more obvious only because it is much closer to the observer. In fact, the further away an object is, the slower it will seem to go. The stars are so remote that to all intents and purposes they do not seem to move at all, unless we use very delicate measuring techniques.

Fig. 1. A one-inch line. If this represents the distance between the Earth and the Sun, then the nearest star will be over 4 miles away.

Now let me give a few facts and figures. I have said that the Sun is 93,000,000 miles from the Earth. Neptune, the outermost of the main planets, is 2,793,000,000 miles from the Sun. (I exclude Pluto, which has a more eccentric path which sometimes brings it inside the orbit of Neptune.) Represent the Earth-Sun distance by one inch (Fig. 1). On this scale, Neptune will be at a distance of $2\frac{1}{2}$ feet. What of the nearest star, naturally excluding the Sun? We find that on this scale its distance will be more than four miles; and most of the bright stars visible at night-time will be more remote still. For instance Rigel, a brilliant star in Orion, will have to be taken out to at least 900 miles from our model Earth.

When dealing with distances of this kind, conventional units such as miles or kilometres are too short to be useful, just as it would be rather cumbersome to measure the distance between London and Paris in inches. We need something better, and fortunately Nature has provided us with a convenient unit.

Light moves at approximately 186,000 miles per second (actually 186,282·3959 miles per second, to within an accuracy of about a yard), and so in one year it can cover 5,880,000,000,000 or rather less than six million million miles. This distance, the astronomer's light-year, is our stellar unit. The nearest of the so-called Fixed Stars is 4·2 light-years away, corresponding to over 24 million million miles; Rigel is about 900 light-years from us. By contrast, light can travel from the Sun to the Earth in only a little over eight minutes, and can leap from the Moon to the Earth in only one and a quarter seconds.

This may do something to show how vast the universe is, and it explains why the stars appear practically motionless. It also means that once we go beyond the Solar System, our knowledge of the universe is bound to be hopelessly out of date. We see Rigel not as it is now, but as it used to be 900 years ago, which takes us back to William the Conqueror. If Rigel were suddenly extinguished, we would know nothing about the disaster until about A.D. 2870. Incidentally, since radio waves travel at the same speed as light (a point to which I will return later) we can forget any ideas of quick interstellar communication, quite apart from the practical difficulties. If anyone living on a planet associated with Rigel could send us a wireless message powerful enough for us to pick up, it would take 900 years to reach us.

When we come back to the Solar System, the distances involved are so much less that we can notice definite motion even over a few hours. This is why the planets wander from one constellation into another; the very word 'planet' is derived from a Greek term meaning 'wanderer'. Though a planet looks superficially very like a star, its movement betrays it, as was first realized a very long time ago.

To be more precise, a planet looks like a star when observed with the naked eye, but once we use optical aid the situation is very different. A planet is close enough to show a definite disk, and there are obvious details to be seen. Venus and Mercury show phases, or apparent changes of shape, because they are closer to the Sun than we are, and everything depends upon how much of their sunlit part happens to be turned in our direction. (Clearly, the Sun can illuminate only half the planet at any one time, so that one hemisphere shines while the other is dark.) Mars displays polar caps and dark markings against its red disk; Jupiter is crossed by cloud belts, while Saturn is distinguished by its superb set of rings. But the immensely remote stars cannot be seen as disks at all. Even the world's largest telescopes show them only as dots of light, and their apparent diameters are too small to be measured directly. If you turn a telescope toward a star and see a large, shimmering ball, you may be sure that there is something wrong with the focusing of the telescope! Therefore, most of our knowledge of the stars has had to be obtained by less direct methods.

At this point it seems worth recalling the famous nursery rhyme which begins 'Twinkle, twinkle, little star . . .' because it is rather more relevant than might be thought. Stars do twinkle; but this effect—known technically as scintillation—has nothing directly to do with the stars themselves. It is due entirely to the Earth's atmosphere, which is unsteady. As the light from the star comes to us through the air, it is distorted, and twinkling results. Note that a star which is high up will twinkle much less than a star which is low over the horizon, because its light is passing through a lesser thickness of atmosphere (Fig. 2). A planet, which appears as a small disk rather than a point, twinkles less than a star, though it is true that a planet very low in the sky can sometimes scintillate quite strongly.

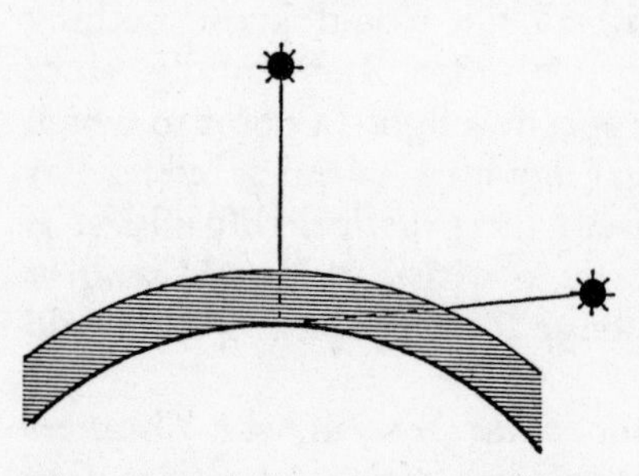

Fig. 2. Star-twinkling. The light from the overhead star is passing through a thinner layer of atmosphere than with the low-down star (to the right of the diagram). Therefore, the overhead star twinkles less.

Apart from some differences in colour and very obvious differences in brightness, one star looks very much the same as another when seen with the naked eye. Telescopically, the variety of the stars becomes more evident. The colours show up well; some stars are orange, while others are distinctly red, though admittedly most of the ordinary stars are white. We see pairs and groups of stars, sometimes with lovely contrasting hues. Then, too, there are the variable stars, which do not shine steadily, but brighten and fade over short periods. Normally, one cannot see a star changing its brightness perceptibly; but the variations become obvious enough after a night or two—in some cases, less. Occasionally a star suffers a tremendous outburst, and flares up to many times its normal brightness in what we call a nova explosion, while even more rarely a star may literally destroy itself in a blaze of radiation which we term a supernova. Also contained in our star-system are patches of dust and gas which we call nebulæ, and inside which we believe that fresh stars are being formed.

The star-system in which we live is known as the Galaxy. It is a large place by any standards; its diameter is of the order of 100,000 light-years, and it is also well-populated, since it contains a grand total of around 100,000 million stars. Of these, our Sun is a perfectly ordinary, undistinguished member, and it seems absurd to suppose that it is unique in being attended by a family of planets. Though we have no proof, most modern astronomers consider that planetary systems are likely to be common, and that life too is probably widespread. The trouble is that so far, at least, we cannot see planets of other stars, though we have excellent evidence that they really do exist.

Even when we consider the Galaxy, we are still dealing with only a small part of the universe. Far away in space—so remote that their light takes millions of years to reach us—we can see other galaxies, each of which contains its own quota of suns. With present-day equipment we can penetrate out to several thousands of millions of light-years, though we cannot yet claim to be able to probe as far as the boundary of the universe (even if the universe is finite in extent, which is very much of an open question).

It has been said, with truth, that we have learned more about the universe in the last twenty years than we had been able to do in the previous two centuries. Space research has something to do with this, since there are some kinds of observations which have to be carried out from above the top of our shielding atmosphere; but much of the work has been done from ground level, and striking advances are being made all the time. Entirely new classes of objects have been found, and things such as quasars, pulsars and black holes are now in the forefront of astronomical research, though as recently as 1962 their existence was completely unsuspected. In spite of this, however, some of the most fundamental of all problems remain to be solved, including what is perhaps the most basic of all: How did the universe begin, and will it ever die?

This outline sketch is very fragmentary, but I hope that it will serve as an introduction. In any case, it may show that our picture of the universe today is very different from anything which our ancestors could have suspected. Yet it would be both unfair and illogical to laugh at the old ideas; every science has had to have a beginning, and it is safe to say that astronomy is the oldest science of all.

Chapter Two

WATCHERS OF THE STARS

We do not know when men first began to study the stars. No doubt the cave-dwellers and the Ice Age hunters looked at them with interest, but without the slightest idea of what they might be. For that matter, the early civilizations were no better informed, and the general impression was that the stars must be small lamps fixed on to a solid crystal sphere which turned around the flat Earth once a day.

This may sound peculiar in A.D. 1974, but it was not at all peculiar in the dim past. There was no obvious reason to doubt that the Earth must be both flat and motionless, and certainly the sky does seem to make one full circuit in a period of roughly 24 hours. The stars appear to travel round the world 'all together', just as specks of mud upon a football will do if the football is spun round, and it was more logical to suppose that the sky moved than to believe that the Earth itself were spinning. It was clear that the Sun and Moon share in this daily motion, though each has an individual motion as well; and at a rather later stage (certainly before Greek times) the individual shifts of the five known planets were also recognized. Early sky-watchers did manage to draw up reasonably accurate calendars, and the length of the year was given as 365 days. To the ancients, it did not much matter whether the Sun went round the Earth or the Earth went round the Sun; the period was 365 days in either case.

The next step was to divide the stars up into groups or constellations. This was done in several countries, and again we cannot be sure which race originally devised the constellations we use today. It may have been the Chaldæans, who came from what is now Mesopotamia; it may have been the Minoans of Ancient Crete, about whom we still do not know nearly as much as we would like. The Chinese had their own groups; so too did the Egyptians, who also made some excellent measurements of the positions of the stars relative to each other.

Everyone associates Egypt with those remarkable structures

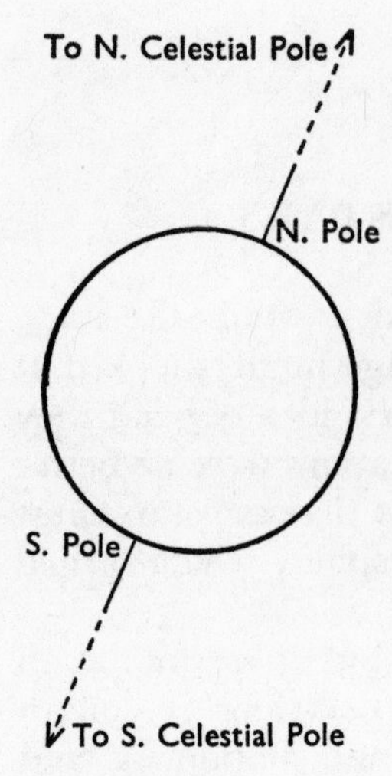

Fig. 3. The celestial poles.

which we call the Pyramids, and there is no doubt at all that these Pyramids were astronomically lined up. Egyptian astronomers knew that everything in the sky seems to move round one particular point, and in their day this point—the North Celestial Pole—lay close to the rather undistinguished star Thuban, in the constellation of the Dragon. The reason for this movement is quite straightforward, and Fig. 3 will make it clear. The Earth spins on its axis; this axis points northward to the celestial pole, and the apparent rotation of the sky is due to the real rotation of the Earth. The Egyptians, of course, put a totally different interpretation upon matters, but there was nothing the matter with their measurements, and when they set up the Great Pyramid they aligned the main gallery inside the structure with respect to the celestial pole. This was lucky for modern researchers, since it gave a clue as to the date when the Pyramid was built—for, oddly enough, the north celestial pole is no longer where it used to be. Nowadays it is not near Thuban, but is close to a brighter star, Polaris in the neighbouring constellation of the Little Bear.

The reason for this shift is that the Earth is not a perfect sphere. The diameter measured through the equator is 7,927 miles, but only 7,900 miles if measured through the poles, so that the equatorial zone bulges out slightly. The Sun and Moon pull upon this bulge, and the direction of the Earth's axis shifts slowly, rather in the manner of a gyroscope which is about to topple (Fig. 4).

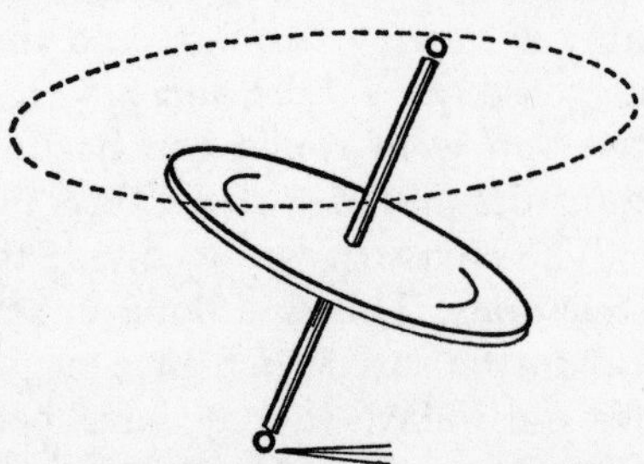

Fig. 4. Precession, as illustrated by a toppling gyroscope. The main difference is that the gyroscope completes one full circle in a second or two, whereas the Earth takes 26,000 years.

There are many complications to be taken into account but, broadly speaking, the celestial pole describes a circle in the sky, taking 25,800 years to complete the full turn. The phenomenon is known as 'precession'. In the time of the Pyramid-builders, the polar point lay near Thuban; today it is near Polaris, and is moving still closer (Fig. 5). Polaris will be at its closest to the actual pole in the year 2012. By A.D. 4000 Alrai in the constellation of Cepheus will occupy the position of honour, while by A.D. 14,000 we will have a really brilliant pole star—Vega in Lyra, the Lyre, the fifth brightest star in the entire sky.

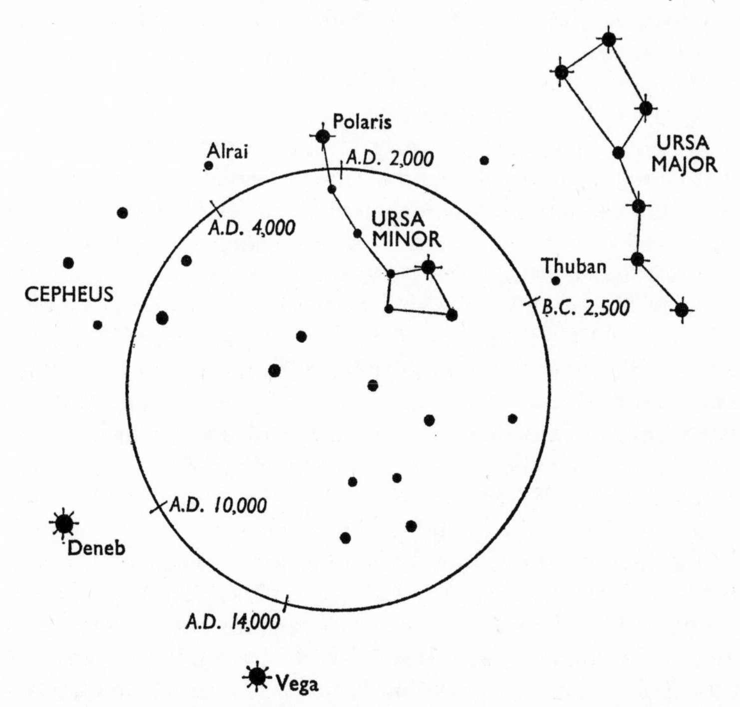

Fig. 5. The changing position of the north celestial pole.

All this shows that the Egyptians were good observers, and that they were accurate in their measurements. They paid great attention to astronomy, and put it to practical use. Their whole economy was founded upon the annual flooding of the Nile, and obviously they needed to know when to expect it. They found that it started at about the date when Sirius, the most brilliant of all the stars, first became visible in the dawn sky, so that they took this 'heliacal rising' of Sirius to mark the beginning of the flood season. This is worth noting, because it shows that even in those far-off times astronomy was not a purely academic study.

Yet the Egyptians never made much progress in theoretical astronomy. They were content to watch, and were reluctant to try to reason out why things happen as they do. The same was true of the Chinese, and real advances date only from the time when the Greeks first came into the picture.

Scientific historians always make a great deal of the so-called 'Greek miracle', and there is no doubt at all that the philosophers of Ancient Greece were responsible for remarkable progress. On the other hand they did not manage it quickly, as is often thought. The early Greeks learned much of their astronomy from Egypt, and their ideas were very primitive. The first of the Ionian philosophers was Thales of Miletus, who was born before 600 B.C.—and was, incidentally, the prototype absent-minded astronomer; according to legend he was once walking along, looking at the stars, when he fell into a well.* The last great astronomer of the Greek school was Ptolemy of Alexandria, who may actually have been Egyptian by birth, and who died in or about A.D. 180. This means that the whole period extended over 800 years; and Ptolemy was as far away from Thales in time as we are from the Crusades.

Thales himself had no doubt that the Earth must be the centre of the universe, and probably he regarded the stars as being fastened on to the vault of heaven. Some of his contemporaries and successors held curious views. One philosopher, Anaximander, believed the stars to be 'compressed portions of air, in the shape of wheels filled with fire, emitting flames at some point from small openings'; another, Xenophanes, taught that 'the stars are made of clouds on fire; extinguished every day, they are rekindled at night like coals; their risings and settings are lightings and extinguishings respectively'. And Heraclitus of Ephesus estimated the diameter of the Sun as one foot—which, as we now know, is something of an underestimate!

* I am no Thales, but I once had a similar experience involving a ditch filled with liquid mud, so that I am very careful not to laugh.

On the credit side, Pythagoras, the great geometer, certainly realized that the Earth is a globe, and the flat-earth theory was finally killed by Aristotle,* who lived from about 384 to 322 B.C. Aristotle was an observer as well as a theorist, and he realized that the idea of a flat world did not fit the facts. For instance, the lovely southern star Canopus can be seen from Alexandria, but never rises from the more northerly latitude of Athens. This can be explained easily enough if we take the Earth to be a globe, but not otherwise.

The next major steps were taken by Aristarchus of Samos (310 to 250 B.C.) and Eratosthenes of Cyrene (276 to 196 B.C.). Aristarchus was bold enough to suggest that the Earth might be nothing more nor less than a world circling the Sun, though unfortunately few people believed him. Eratosthenes is best remembered for his ingenious method of measuring the size of the Earth, which led him on to an amazingly accurate result. By this time there were at least some glimmerings of knowledge about the make-up of the universe, and it had become fairly generally assumed that the Moon and planets were 'earthy' while the Sun must be blazing hot; but almost nothing was known about the stars, and most of the Greek philosophers continued to believe in the idea of lights tacked on to an invisible, transparent sphere.

About 150 years before Christ, when Greek learning was still at its height, there flourished a mathematician named Hipparchus. We know practically nothing about his life, and neither have we any of his original writings, all of which have been lost; but we do know that he spent many years in drawing up a star catalogue. Considering that he had to develop and build his own instruments, his accuracy was remarkable. He was also the first man to make proper use of trigonometry, and it was he, too, who discovered precession—that is to say, the wandering of the celestial pole.

Finally, in our brief sketch of classical astronomy, we come to Claudius Ptolemæus, better known to us as Ptolemy. His

* Or nearly so. The International Flat Earth Society still exists, though the results from artificial satellites and space-probes have caused its membership to fall rather alarmingly. Its members believe that the world is shaped like a gramophone record, with the North Pole in the middle and a wall of ice all round; when asked what happens underneath, they tend to be a little evasive. I have discussed them fully in my book *Can You Speak Venusian?* (1972) and further comments here would be out of place!

personality, like that of Hipparchus, is unknown to us, and there seems to be no chance of our finding out anything now, but of his skill there can be no doubt at all. Using Hipparchus' work as a basis, he compiled a star catalogue which remained the best for many centuries. He was also a geographer, and it was Ptolemy who drew up the first world map which was based upon scientific principles rather than hopeful guesswork. Fortunately for posterity, he wrote a book which summarized most of what ancient science had found out. The original has not survived, but the Arab translation has come down to us under the name of the *Almagest*, and its value is incalculable. Without it, we would know far less about science of the Classical period than we actually do. We owe a great debt to Ptolemy, who was well nicknamed 'the Prince of Astronomers'.

It is true that Ptolemy had no faith in the idea of a moving Earth, and he still believed that the sky must turn round us once a day; but he did his best to work out a theory which would fit the facts—and he succeeded, even though the result was horribly awkward and cumbersome. Briefly, he thought that the Sun, Moon and planets revolved round the Earth at different distances, starting with the Moon and ending with the furthest known planet, Saturn. Well beyond Saturn lay the sphere of the fixed stars. Ptolemy did not pretend to know how far away the stars were, but he did realize that they must be very remote indeed.

Before going on with the story, I must, I think, pause in order to say something more about Ptolemy's star-catalogue. In it he gave a list of 48 constellations, all of which are still to be found in our modern maps even though the details and the boundaries have been altered. Of course, Ptolemy knew that the constellation patterns were to all intents and purposes unchanging, but he could not cover the whole sky, because so far as we know he spent his whole life in Alexandria, and never saw the stars of the far south.

There are some striking star-groups in the sky. Most people can recognize the outline of the constellation which is known officially as Ursa Major (the Great Bear), though many people know it as the Plough or the Big Dipper. Its seven stars make up a distinctive pattern, and once it has been recognized it is not likely to be forgotten. Ptolemy probably thought that the stars in the Bear made up a true system, because he knew nothing about the nature of the stars—and if each one were a lamp fixed on to a crystal sphere, it followed that all the stars would have to lie at the same distance from us. Actually, this is not so, and the Great Bear provides an excellent example. Of the two stars in the 'tail', Mizar and Alkaid, it has been found that Mizar is 88 light-years from us, while Alkaid is 210 light-years away (Fig. 6). This means that Alkaid is much further away from Mizar than we are, and the two appear to lie side by side simply because they are in much the same direction as seen from Earth. If we were observing from a position between the two, Mizar and Alkaid would be on opposite sides of the sky.

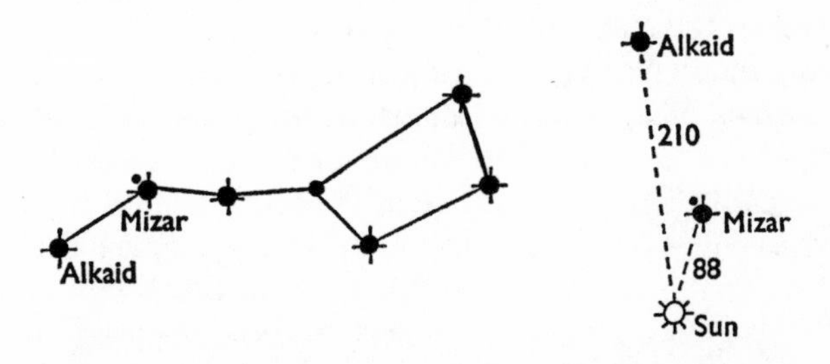

Fig. 6. (Left) the Great Bear, with Alkaid and Mizar apparently side by side. (Right) The real relative distances; Alkaid is almost three times as far away.

Though the constellation patterns have no real significance, they are both convenient and spectacular; and they have been given names, drawn chiefly from ancient mythology. It has even been said that the sky is a complete picture-book. Most of the main groups are associated with old tales; and some of them are fascinating. For instance, there is the legend of Perseus and Andromeda, which is unusual in having a happy ending. We are told that Queen Cassiopeia boasted that her daughter Andromeda was more beautiful than the sea nymphs—thereby offending Neptune, god of the oceans, who sent a monster to ravage the kingdom. In despair Cepheus, the King, was forced to offer Andromeda as a sacrifice. The luckless maiden was chained to a rock to await the monster's arrival, but was saved in the nick of time by the hero Perseus, who had been on an expedition to kill Medusa, an unpleasant lady with snakes instead of hair. As he sped by, on his magical winged shoes, Perseus saw Andromeda chained to the rock. With true chivalry he descended, introduced himself, and then turned the sea-monster to stone by the simple expedient of showing it Medusa's head, after which it was only a question of time before wedding-bells rang out.

Look into the sky, and you will find most of the characters in the legend. Perseus is prominent, clutching Medusa's head; Andromeda is next to him: Cepheus is admittedly rather obscure, but the five stars making up the W-pattern of Cassiopeia are familiar to everyone interested in the sky. Even the sea-monster Cetus is there, though in some maps it is relegated to the status of a harmless whale.

(Note, incidentally, that in astronomy we always use the Latin names of the constellations: Ursa Major instead of the Great Bear, Leo instead of the Lion, Boötes instead of the Herdsman, and so on. There is nothing difficult about these names, and it is best to become used to them at once. I have given a full list, with English equivalents, in the Appendix.)

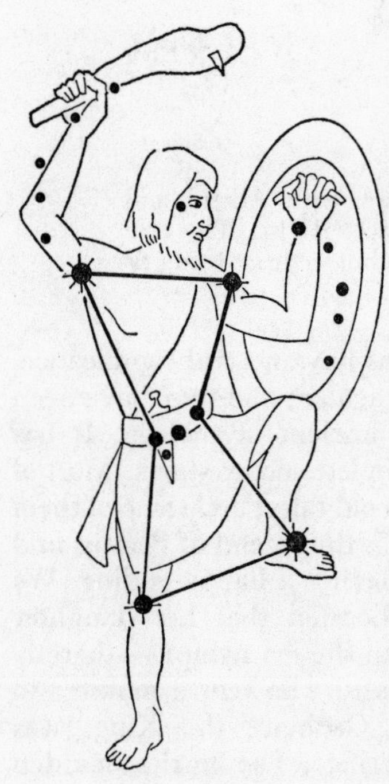

Fig. 7. Orion.

Among other mythological characters we have Orion, the Huntsman, who boasted that he could kill any creature on earth, but was eventually dispatched by being stung by an unfriendly scorpion. The scorpion is also in the sky, but has been placed directly opposite to Orion (Fig. 7), so that the two can never meet. Then we have Pegasus, the Winged Horse; Hercules, the great hero, together with some of his victims such as Leo; and of course the two Bears, whose legend is rather less well known. It is said that Callisto, the daughter of King Lycaon of Arcadia, grew up to be lovelier than Juno, the queen of Olympus, and in jealousy Juno turned the unfortunate princess into a bear. Years later, Callisto's son Arcas encountered the bear, and was about to shoot it when Jupiter, the king of the gods, intervened. He turned Arcas into a bear also, and then caught both animals by their tails, swinging them up into the sky and granting them immortality. This is why both the Great Bear (Callisto) and the Little Bear (Arcas) have tails of decidedly un-ursine length!

It is tempting to dwell upon these fascinating old stories, but we must remember that while we regard them as fantasy our ancestors took them very seriously. Moreover, there was also the question of astrology, which set out to link the star-patterns and the positions of the planets with human characters and destinies. There are still some ignorant people who confuse astrology with astronomy, and so I must say a little about it here.

The most important constellations in astrology were those of the Zodiac. This is the belt around the sky in which the Sun, Moon and bright planets are always to be found. The explanation is quite straightforward. The planets move round the Sun in roughly the same plane, so that if we draw a map of the Solar System on a flat piece of paper we are not very far wrong. Of the planets visible to the naked eye, only Mercury has an orbit whose plane is tilted to that of the Earth by more than 5 degrees; and therefore we see the planets only in certain directions against the starry background. The twelve Zodiacal constellations form a band right round the sky, beginning with Aries (the Ram) and ending with Pisces (the Fishes). Astrologers claimed that the positions of the Sun, Moon and planets in the Zodiac at the time of a baby's birth would affect the whole of his (or her) life. For instance, I happen to have been born on March 4, when the Sun was in the astrological sign of Pisces; astrologers will say that my character would have been different if I had been born on—say—April 4, with the Sun in Aries!

There is no scientific basis for astrology, and nobody with any real intelligence takes it seriously nowadays, even though it is often amusing to read the 'What the Stars Foretell' columns in the less demanding Sunday newspapers. Yet up to a few centuries ago astrology was regarded as a true science, and astronomers were also astrologers. Ptolemy was no exception, and to him it is likely that the constellations were rather more than chance patterns in the sky.

Another feature of Ptolemy's catalogue was that he divided the stars into grades or 'magnitudes' of apparent brilliancy. The system was not Ptolemy's own, but he was wise enough to adopt it. The general scheme is to class bright stars as being of magnitude 1, fainter stars of magnitude 2, and so on down to magnitude 6, which includes the dimmest stars visible to the naked eye under good conditions. The smaller the magnitude, the brighter the star; the scheme works rather in the manner of a golfer's handicap, with the most brilliant performers having the lowest numerical values. We can even take the comparison further. Really outstanding golfers have handicaps which are lower than zero (scratch), and similarly the most prominent stars have zero or negative magnitudes; Rigel in Orion is +0·1, while Sirius, the brightest star of all, is minus 1·4. After telescopes were invented, in the seventeenth century, stars fainter than the sixth magnitude could be recorded, and nowadays our giant instruments can reach down to magnitude +23—though I am afraid that not even the world's largest telescope can attain a magnitude equal to my own golf handicap of +36!

As I have already pointed out, Ptolemy spent all his active life in Alexandria, where the latitude is 31 degrees North. This meant that there were certain stars which he could not see, because they lie too far south in the sky and never rise above the Alexandrian horizon. And this brings us back to Pythagoras' original proof that the Earth is a globe instead of being as flat as a pancake.

It had long been known that the appearance of the sky varies according to the position of the observer, and anyone who has done much travelling will know what is meant. In Britain the north celestial pole is fairly high up, so that groups near it, such as the Great Bear, never set; they are 'circumpolar', and remain above the horizon all the time, so that they are always to be seen whenever the sky is clear and dark. Stars further from the pole, such as the brilliant orange Arcturus, are not circumpolar, since for a part of their 24-hour sweep

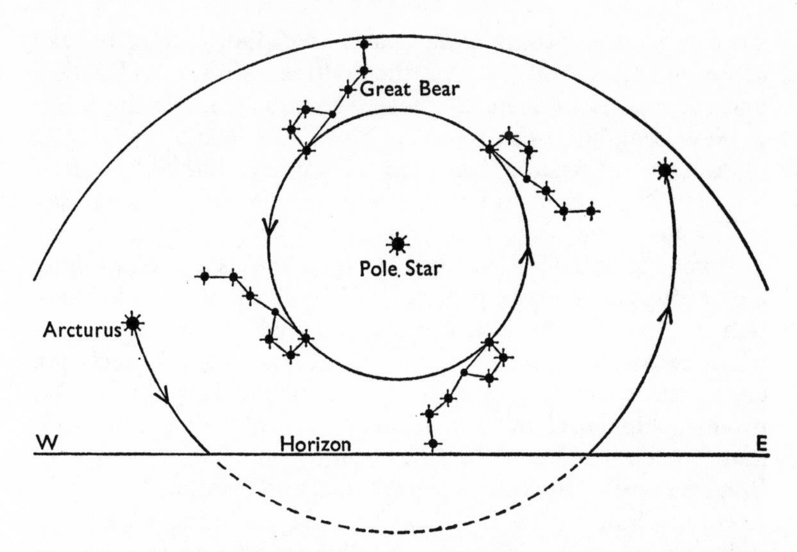

Fig. 8. Circumpolar and non-circumpolar stars. From England, the Great Bear never sets, but Arcturus does.

round the pole they are below the horizon. Fig. 8 shows what is meant.

The further north we travel, the higher the Pole Star will become, and an observer standing at the north pole of the Earth will see Polaris straight overhead, at the 'zenith'.* In fact, the altitude of the celestial pole above the horizon is equal to the observer's latitude on the Earth. If you are in London, where the latitude is 51½°N., Polaris will be 51½ degrees above the horizon; from Oslo (latitude 60°N.) Polaris will be 60 degrees up, and so on. The latitude of the north pole is 90 degrees, so that from here this is also the altitude of Polaris; in other words, it is directly overhead, with the celestial equator lying round the horizon.

If we move toward the equator, Polaris sinks in the sky. From Alexandria it is only 31 degrees above the horizon, and the Great Bear is no longer circumpolar. By the time we reach the equator, our latitude is 0 degrees, and so Polaris has an altitude of 0 degrees, which means that it lies on the horizon; from latitudes south of the equator it can never be seen at all.

* Or virtually so. To be precise, Polaris is almost a degree away from the polar point, so that a slight correction is needed.

To compensate, groups from nearer the south pole of the sky come into view, and we meet the Southern Cross, the Centaur and the rest. Londoners can never see these constellations, just as New Zealanders can never see the Great Bear.

Ptolemy's sky-map, then, was incomplete. For that matter, his 48 constellations did not even cover the whole of the sky which he could see, since some areas were left out. Later astronomers extended the list, even to the extent of forming new groups by chopping pieces off the original 48, and there was a time—in the seventeenth and eighteenth centuries—when constellation-forming became all the rage. The climax came with a German of high repute, Johann Elert Bode, who produced maps of the sky which included new groups with names such as Sceptrum Brandenburgicum (the Sceptre of Brandenburg), Officina Typographica (the Printing Press) and Lochium Funis (the Log Line), to say nothing of Globus Aërostaticus (the Balloon). At this point astronomers in general decided to call a halt, and to drop some of the smaller constellations with barbarous names. The modern list has been reduced to 88 separate groups.

It is idle to pretend that the constellation patterns are either logical or convenient. It has even been said that they seem to have been designed so as to cause as much confusion as possible. However, they have become so ingrained that they will certainly not be altered now, and occasional suggestions for drastic revision have been firmly turned down. (The worst idea yet was to re-name the constellations in honour of political leaders. One can well imagine the international crises which would ensue if, for instance, a brilliant star were claimed by both Mr. Brezhnev and Mr. Ian Smith!)

Because the constellations round the south pole of the sky are modern, inasmuch as they do not date back beyond the seventeenth century, they are in general smaller and more numerous than those of the far north. The exception is Argo Navis, originally named in honour of the ship which carried Jason and his companions on their quest of the Golden Fleece. Argo was so huge that it was frankly unwieldy, and now, by order of the International Astronomical Union, it has been unceremoniously broken up into a keel, sails and poop. At the other end of the scale we have Crux Australis, the Southern

Cross, which is actually the smallest constellation in the whole sky, though it contains a high concentration of bright stars. Oddly enough, the Cross was not a separate entity until 1679, when it was formed by an otherwise obscure astronomer named Royer. Previously it had been included in the much larger constellation of Centaurus, the Centaur.

To return to Ptolemy: he left astronomy in a state of good order, but he had no immediate successors. During the so-called Dark Ages which followed the collapse of the Roman Empire, science was more or less at a standstill. When astronomy came to the fore again, it was by way of astrology. A thousand years ago the Arabs were busy drawing up star catalogues which were more accurate and detailed than Ptolemy's, and it was mainly the Arabs who gave the bright stars the individual names by which we still know them. Names such as Betelgeux, Rigel, Altair and so on are of Arabic origin.

The Arabs needed their star catalogues, together with tables giving the apparent movements of the planets, in order to draw up astrological horoscopes; but navigation also benefited. As travel became more and more extensive, there was increasing use of 'steering by the stars', a trend which increased when astronomy spread from the Arab countries back into Europe. Then, in the last part of the sixteenth century, came a star catalogue which was far better than anything previously attempted.

The man responsible was Tycho Brahe, a Dane who was undoubtedly one of the oddest and most colourful characters in the history of science. Tycho took an interest in astronomy from childhood, but his career really began in 1572, when he discovered a brilliant new star in the northern constellation of Cassiopeia. The star was spectacular. Indeed, it was so bright that it could be seen with the naked eye in broad daylight, and we now know that it was what is called a supernova, marking the destruction of a massive star. I will have more to say about it, and about Tycho himself, in Chapter 12. Meanwhile, it is enough to note that between 1576 and 1596 Tycho worked away at his island observatory of Hven, in the Baltic, compiling an amazingly good catalogue of naked-eye stars and making excellent measurements of the wanderings of the planets. On his death, in 1601, his observations came into the possession of his last assistant, Johannes Kepler, who used them to give decisive proof that the Earth moves round the Sun instead of lying at rest in the centre of the universe. Ironically, this was something which the imperious and astrologically-minded Tycho could never have brought himself to believe.

Yet another catalogue was produced by a Bavarian, Johann Bayer. It was published in 1603, and was important because it introduced the modern system of star nomenclature.

Giving stars individual names, as the Arabs had done, was all very well, but it was hardly convenient, particularly as some of the names were decidedly tongue-twisting; typical examples were Zubenelgrnubi, Azelfafage, Alkaffaljidhina, and Dschubba. Also, it would be out of the question to name every star in the sky. Bayer's method was to take each constellation and give its stars Greek letters. Ideally, the brightest star in each group would be Alpha, the second brightest Beta, and so on down to Omega, the last letter in the Greek alphabet. In this case the leader of Ursa Major would become Alpha Ursæ Majoris (Alpha of the Great Bear), though it would still retain its Arab proper name of Dubhe. As is usually the case, the rule was not strictly followed, and we have numerous departures from it; in Ursa Major itself, for example, the brightest star is Epsilon, followed by Eta, Alpha, Zeta, Beta and then Gamma. However, the scheme was so logical that everyone adopted Bayer's lettering. Since the Greek alphabet is always used, it will be helpful to give a full list.

α Alpha	η Eta	ν Nu	τ Tau
β Beta	θ Theta	ξ Xi	υ Upsilon
γ Gamma	ι Iota	ο Omicron	φ Phi
δ Delta	κ Kappa	π Pi	χ Chi
ε Epsilon	λ Lambda	ρ Rho	ψ Psi
ζ Zeta	μ Mu	σ Sigma	ω Omega

Nowadays, the old proper names are commonly used for only the twenty or thirty brightest stars in the sky, plus a few fainter ones of special interest. The rest have become virtually obsolete, though they still crop up occasionally.

With Tycho and Bayer, we come to the end of the first period of stellar astronomy. It was almost purely 'positional', and all that could be said about the stars themselves was that they were a very long way away. However, a new era was at hand. In the first decade of the seventeenth century we come to the introduction of the telescope, which has led on to our finding out not only how the stars move, but what they really are.

Chapter Three

THE STELLAR SKY

According to some authorities Nero, who has the doubtful distinction of being the worst emperor whom even Rome had to suffer, wore a quartz monocle. Whether this is true or false seems uncertain; but we may be sure that aids to sight such as spectacles were not known in Classical times. Yet spectacles had come into use by the beginning of the sixteenth century, and it is rather curious that the principle of the telescope was not discovered until considerably later.

So far as we know, the honour goes to Hans Lippershey, of Middelburg in Holland. It is possible that he had been anticipated, but he certainly built a telescope in 1608, and the news of his invention spread as quickly as any news could do in that era of slow, unreliable communication. Galileo, professor of mathematics at the University of Pisa, heard about it in 1609, and, "sparing neither trouble nor expense", proceeded to make a telescope for himself. It was much less effective than modern binoculars, but it was good enough for Galileo to make a series of spectacular discoveries. He saw the mountains and craters of the Moon, the satellites of Jupiter, the phases of Venus, and other wonders which had hitherto been quite unknown. When he turned his telescope toward the stellar sky, he found that the luminous band known as the Milky Way is made up of countless faint stars.

The telescopes built by Galileo and his contemporaries were of the refracting type. The light from the object to be studied is collected by a glass lens or object-glass, and the image is magnified by a second lens, known as the eyepiece (Fig. 9). (The reflecting telescope, in which the light is collected by a mirror instead of an object-glass, did not come upon the scene until about 1670, when Isaac Newton developed it.) For various reasons which need not concern us at the moment, early refractors produced an unwelcome amount of false colour, so that any bright star would appear to be surrounded by rings of gaudy hues which had no real existence. One method of reducing the trouble was to make the telescopes very long, so that in some cases the object-glass had to be fixed to a mast.

Such an 'aerial telescope' was used by Hevelius, who lived in Danzig (the modern Gdańsk). He began his active observing career in 1644, and his main telescope had an object-glass ninety feet away from the eyepiece. It was, to put it mildly, awkward to use, and I have never understood how it could have been made effective. Nevertheless, Hevelius carried out excellent work, notably in mapping the Moon. He also drew up a star catalogue, though he was never able to bring himself to use telescopic sights. Hevelius' catalogue included over 1,500 stars; and considering the difficulties under which he worked, his positional measurements were very good.

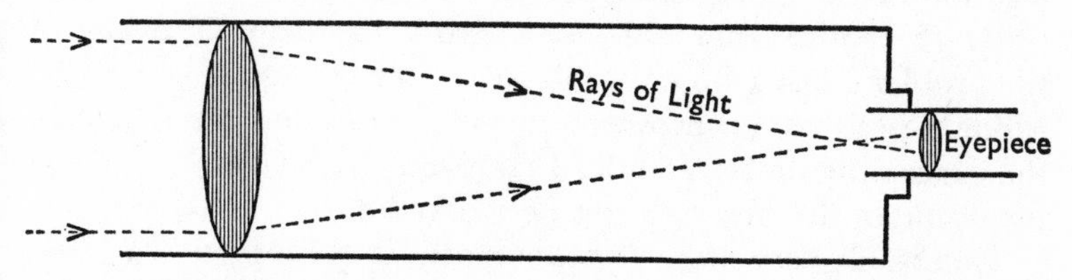

Fig. 9. Principle of the refractor. The light is collected by an object-glass, and the image is magnified by means of an eyepiece.

The whole question of star cataloguing came to a head in the reign of King Charles II, not because of any concern with pure science, but for strictly practical reasons. England has always been a seafaring country, and in those days the only way in which a sailor could find his position, when far out to sea, was by means of the Sun, Moon and stars—in fact, by what we now term astro-navigation. Latitude is easy to find, since all that has to be done is to measure the altitude of Polaris and make a few corrections. (Things are more troublesome in the southern hemisphere, with no Polaris and no bright south polar star, but the principle is just the same.) Longitude is much more troublesome, and it was therefore of more than academic interest when a Frenchman named Saint-Pierre suggested to English scientists that there could be a quick, reliable way of longitude-finding by making observations of the Moon and the stars.

Actually, Saint-Pierre's method was not practicable, but the

King wisely appointed a scientific committee to investigate the whole problem. He was able to call upon men of great ability; the Royal Society had been founded, and among the outstanding leaders of the time were John Flamsteed, Edmond Halley, Christopher Wren (who, incidentally, was professor of astronomy at Oxford before making his name as an architect), and of course Newton himself.

The Rev. John Flamsteed, already widely known as a careful and painstaking observer, strongly supported the idea. In brief, it involved measuring the position of the Moon against the stars. The Moon is so close to us that it moves relatively quickly across the sky, and so we may regard the sky as a clock face, with the Moon playing the part of a moving hand. The position of the clock-hand, i.e. the Moon, therefore gives the time. (I realize that this is a gross over-simplification of the problem, but I hope it will suffice for the moment.) If the Moon's position is measured really accurately, we can find the exact time on the standard meridian of the Earth, and the longitude of the observer can be worked out.

This is all very well, but obviously it means knowing the star-positions very accurately, to say nothing of the precise way in which the Moon behaves. Tycho's star catalogue, already almost a century old, was not good enough, because it had been made before the invention of the telescope. The only solution, then, was to compile a better one. King Charles decreed that the stars must be "anew observed, examined and corrected" for the use of British seamen, and for this purpose a full-scale observatory was set up in Greenwich Park.

It cannot be said that the early history of the Royal Observatory was entirely smooth. Typically, Charles raised the money for the original building by the sale of "old and decayed" gunpowder, after which he commissioned Wren to undertake the design. The Observatory was completed in 1675, and Flamsteed was put in charge of it. However, the King's generosity did not extend to providing Flamsteed with any instruments, and the first Astronomer Royal had to fend for himself. His salary, £100 a year, was hardly princely, and altogether he worked under difficulties. For a time his only official assistant was a "silly, surly labourer" (to use Flamsteed's own description) who rejoiced in the name of Cuthbert. The arrangement was not wholly successful. Cuthbert, at a salary of £26 a year, was too inclined to remove himself to the local inn at times when he was needed to help the Astronomer Royal with his observations!

In spite of all this, work on the star catalogue proceeded. Flamsteed was in no real hurry; he was a perfectionist and he was also temperamental, so that he frequently quarrelled with his contemporaries—including Newton and Halley. Then there was trouble with one of the instruments, made by Robert Hooke. All in all, years passed by with Flamsteed working away and still refusing to publish his results. At last, in 1704, Prince George of Denmark, husband of the new sovereign Queen Anne, offered to pay for the publication of the catalogue, and Flamsteed handed a manuscript copy to a Royal Society committee headed by Newton. He added, however, that the catalogue was not yet in its final form, and stipulated that it should not be printed until everything had been re-checked, though the actual observations could go forward.

For a number of reasons, including the death of Prince George in 1708, printing was held up, and still Flamsteed did not produce the final version of his actual catalogue, for which all his fellow astronomers were waiting. Finally, in 1711, the Committee published a large book containing not only Flamsteed's observations, which he had passed for publication, but also the catalogue, which he had not. To make matters worse, the catalogue contained errors, plus a preface, written by Halley, which could not be anything but harmful to Flamsteed's reputation.

A veil is best drawn over the undignified squabble which followed. None of those concerned emerge with much credit, and when Flamsteed managed to lay his hands on a large number of copies of the book he burned them publicly, "that none might remain to show the ingratitude of two of his countrymen". Then, in 1719, he died—and his widow swept into the observatory like an east wind, removing all the instruments, which were legally hers! Halley, the next Astronomer Royal, had to make a fresh start. Yet it is satisfactory to record that Flamsteed's final catalogue was completed in the end by his last assistants, Crosthwait and Sharp. It made its appearance in 1725, and was just as good as had been hoped. Of course its accuracy is low by modern standards, when we have sophisticated instruments and also the advantage of photography; but Flamsteed's catalogue was a tremendous achievement, and a great tribute to the skill and industry of the first Astronomer Royal.*

All the first major star-catalogues, including Ptolemy's, had been made by observers living in the northern hemisphere, and the stars of the far south were inevitably neglected. The first man to make a serious effort to chart them accurately was Edmond Halley, long before he became Astronomer Royal. In 1676 he went to the island of St. Helena, where he remained for long enough to catalogue 360 important southern stars. Halley, best remembered for his studies of the comet which bears his name and which will be bright once more in 1986, was very different in character from the temperamental Flamsteed or the ill-tempered Hooke. He was 63 years old when he became Astronomer Royal, and showed his natural optimism at once by embarking upon a series of observations of the Moon's position which he knew would take him nearly twenty years. It is pleasant to add that he completed the main task before his death in 1742.

One of Halley's greatest contributions to stellar astronomy was his discovery that the three bright stars Sirius, Procyon and Arcturus were not quite where they had been in the time of Ptolemy; and this brings us back to the problem of the individual or proper motions of the stars.

As I have already pointed out, the stars seem to stay in almost fixed positions with respect to each other simply because they are so far away (remember the analogy of the sparrow and the jet). The naked-eye view of the constellations does not alter over long periods. Yet given sufficient time, these tiny proper

* This is not the place to follow through the story of navigation and the stars; it is enough to say that, ironically, the method of 'lunar distances' was never widely used, because it had already been superseded when it became practicable! Up to 1971 the post of Astronomer Royal remained at Greenwich, though the observatory itself was moved to the clear skies of Herstmonceux, in Sussex, during the 1950s, and Wren's original building is now an astronomical museum. Following the retirement of Sir Richard Woolley as Astronomer Royal, in December 1971, there was something of a gap; then Professor Sir Martin Ryle, the great radio astronomer, accepted the post—but as he remained at Cambridge, the connection with Greenwich was officially severed. For a brief period Mrs. Margaret Burbidge became Director of the Royal Observatory, but soon resigned, to be succeeded in 1973 by Dr. Alan Hunter, former Chief Assistant there.

motions build up sufficiently to become noticeable, and to show what is meant let us look at the seven Plough stars of Ursa Major, the Great Bear, in the past, present and future.

Six of the stars are of about the second magnitude, with the other (Megrez, or Delta Ursæ Majoris) distinctly fainter. Most people know the familiar Plough or Big Dipper shape, shown in the middle diagram (Fig. 10). From the upper and lower diagrams of the constellations, it will be seen that while five of the stars are going much the same way, the remaining two—Alpha (Dubhe) and Eta (Alkaid)—are not. Two hundred thousand years ago the pattern would have been as shown in the upper diagram. At an equal period in the future, the Plough-pattern will be equally alien. Orion, too, will alter eventually. Yet all proper motions are very slight. In general, the nearest stars seem to move the fastest, but even the greyhound of the stellar skies, Barnard's Star (of which more anon) takes nearly two centuries to creep across the sky by an amount equal to the apparent diameter of the full moon. Sirius, Procyon and Arcturus move more slowly still, and so Halley's discovery of their proper motions was a major feat.

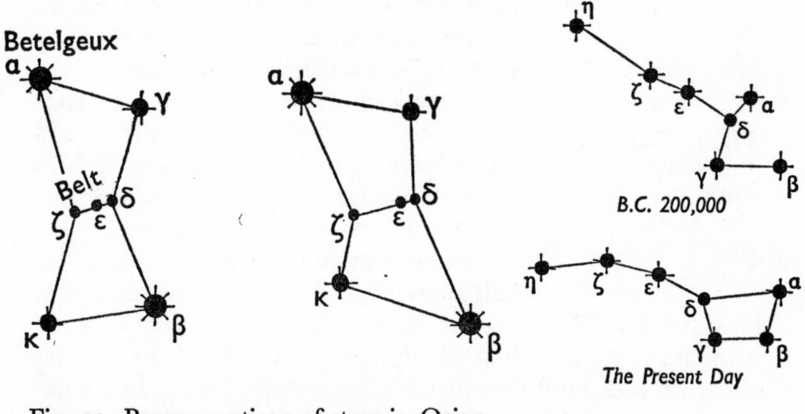

Fig. 10. Proper motions of stars in Orion (*above*) and the Great Bear (*right*). Over a sufficiently long period, both these famous groups will lose their now-familiar shapes.

Of course, these individual motions of the stars have nothing to do with the apparent shift of the celestial pole. The movement of the pole—precession—is due to a real shift in the direction of the Earth's axis, and has no direct connection with the stars themselves. Polaris is not genuinely important; it merely happens to lie in a special position in the sky, though it will not always do so.

If the positions of the stars are to be measured, we must have a standard of reference of the same kind as latitude and longitude on the Earth's surface. Terrestrial latitude is the angular distance north or south of the equator as measured from the centre of the Earth; for instance, the latitude of my home at Selsey is N.50°44′, as shown in the diagram (Fig. 11). Longitude is the angular distance east or west of the Greenwich meridian, also measured from the centre of the Earth. The equator is a 'natural' feature, so that latitudes may be reckoned from it, but there is nothing comparable as a zero for longitude. The only reason why we use the Greenwich meridian is that it was selected by international agreement.

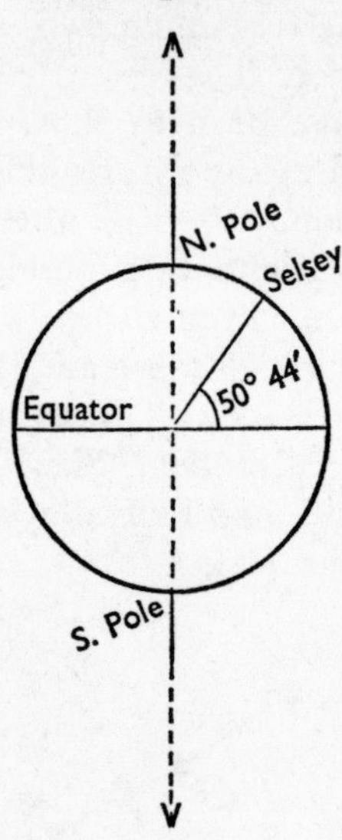

Fig. 11. Latitude on the Earth: 0° for the equator, 90° for the poles. Selsey, shown here, is at latitude N. 50° 44′.

A 'great circle' is a curve which runs right round the surface of the Earth, and whose plane passes through the Earth's centre. In other words, if you could slice the Earth through a great circle you would cut the globe exactly in half. The only parallel of latitude which is also a great circle is, of course, the equator. The Greenwich or prime meridian is the meridian which passes through the north pole, the south pole, and Greenwich Observatory. By the late nineteenth century it had become obvious that longitudes would have to be standardized, and Greenwich was regarded as the timekeeping headquarters of the world, so that a conference was called and agreement was reached. International co-operation was much easier to obtain then than it is now, and the only two nations to demur were—predictably—Ireland and

France. For many years the French reckoned their longitudes from the meridian of Paris, but nowadays everyone accepts that longitude 0° passes through the instrument known as the Airy Transit Circle, set up over a hundred years ago at Old Greenwich by the then Astronomer Royal, Sir George Airy. It is still on view, and you can see it any time you like. (A favourite pastime among sightseers is to be photographed straddling the meridian line, so that one foot is in the eastern hemisphere of the world and the other in the western.)

For the sky, we have to find an equator, and also some equivalent of the Greenwich meridian. It is convenient to imagine that the sky is solid, so that lines can be drawn on the 'celestial sphere'. If we project the plane of the Earth's equator on to this sphere, we mark out the celestial equator, as shown in Fig. 12. We can then measure the angular distance of any star north or south of the equator, which gives the Declination —minus 16°39′ in the case of Sirius, for example, which lies south of the equator. Obviously, the declination of the north celestial pole is +90°; Polaris has a declination of +89°2′, and

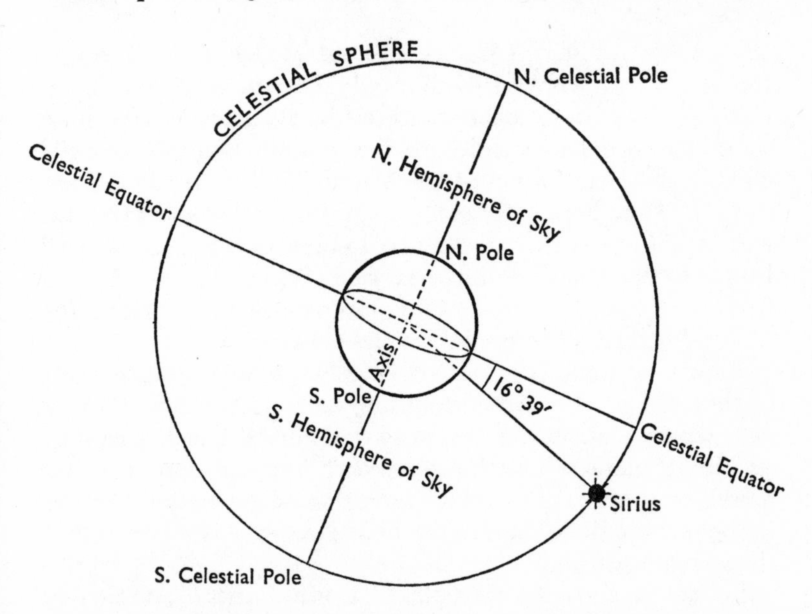

Fig. 12. The celestial sphere, showing the celestial poles and equator. Sirius has a declination of 16° 39′S.

so is less than a degree away. (To be really precise, corrections have to be made so that the measured angles are referred to an imaginary observer who is placed at the centre of the Earth.)

For east-west reckoning we turn first to the Ecliptic, which is defined by the plane of the Earth's orbit. To all intents and purposes this is also the apparent yearly path of the Sun among the stars. The plane of the Earth's equator is tilted with respect to the plane of the orbit at an angle of 23½ degrees (Fig. 13), and this means that the angle between the celestial equator and

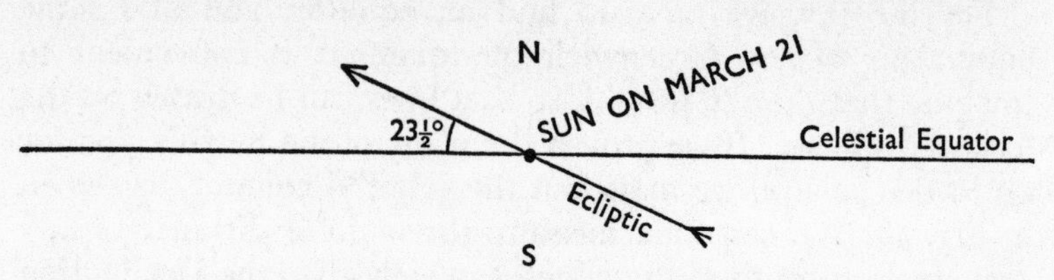

Fig. 13. The angle between the celestial equator and the ecliptic is 23½°. About March 21 each year the Sun crosses the equator, moving from south to north. The point of intersection is called the Vernal Equinox, or First Point of Aries.

the ecliptic is also 23½ degrees. About March 21 each year—the exact date varies slightly, owing to the vagaries of our calendar—the Sun reaches the celestial equator, moving from south to north; this is where the ecliptic and the celestial equator cross, and is called the Vernal Equinox or First Point of Aries. Here is the source for our prime meridian of the sky, and we take it as marking the zero point, so that it could well be nicknamed 'the celestial Greenwich'. Star positions measured from it in an eastward or anti-clockwise direction along the celestial equator give the star's Right Ascension.

Right Ascension is measured in hours, minutes and seconds of time. This may sound confusing at first, but actually it is very convenient. A star is said to 'culminate' when it reaches its highest point above the observer's horizon, and is on his meridian (Fig. 14); the right ascension is the time-difference between the culmination of the First Point of Aries and that of the star concerned.

Let me give you an example. Since the Earth takes one day to spin on its axis, the First Point of Aries is bound to culminate each day; often it does so when the Sun is above the horizon, but this makes no difference to the general argument. Sirius culminates 6 hours 43 minutes after the First Point of Aries has done so, and this means that its right ascension is 6h 43m. Vega, the brilliant bluish star in the constellation of Lyra, culminates 18h 35m after the First Point, and so on.

This First Point is so named because it used to lie in the constellation of Aries, the Ram. Because of precession, it has now moved out into the next-door constellation of Pisces, the Fishes, but the name has not been changed; and doubtless it will still be kept when the equinox moves out of Pisces into the adjacent group of Aquarius, the Water-Bearer. Moreover, we still regard Aries as the first constellation of the Zodiac, though technically it ought to be the second.

Precession, of course, alters the right ascensions of the stars, and in any list you will note that the year or 'epoch' is given. The values for Sirius and Vega quoted above are for epoch 1950, but by now they are slightly different.

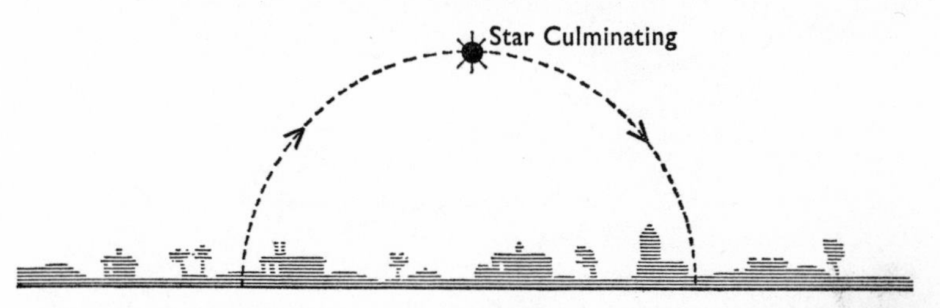

Fig. 14. Culmination of a star: the altitude above the horizon is at its greatest.

All large telescopes are equipped with setting circles, graduated according to right ascension and declination. If you know the co-ordinates for any particular object, all you have to do is to make some quick calculations, set the circles by swinging the telescope, and look through the eyepiece. If the adjustments are correct, and you have made no mistake, the object will be in the field of view. Of course the object is moving across the sky all the time, because of the Earth's rotation, and the telescope has to be fitted with a clock drive which moves it slowly round so as to keep the object in view.

I am afraid that I have digressed rather widely from my historical survey, but the whole question of star positions is so important that it cannot be glossed over, since precise measurement is essential in modern stellar astronomy. For that matter the positions of the Sun, Moon and planets can also be given on the same system, though since they wander around the sky their right ascensions and declinations alter quickly.

Many star-catalogues have been drawn up since Flamsteed's day, and modern accuracy is remarkable. Of course, the development of photography has made all the difference. Meanwhile, let us turn to another fundamental problem—that of the distances of the stars.

Chapter Four

THE DISTANCES OF THE STARS

FLAMSTEED, THE FIRST Astronomer Royal, catalogued the stars. His successor, Halley, recognized proper motions, while in 1728 the man who was later to become the third holder of the office, the Rev. James Bradley, decided to attack a still greater problem, that of finding out how remote the stars really are. The method he proposed to use was that of 'parallax'.

The principle involved is quite simple, and once again we have an everyday analogy. Suppose that a surveyor wants to measure the distance between a small marker-post, P in the diagram (Fig. 15), and a tower T on the far side of a river. Also suppose that he has no boat, and is in no mood for a swim, so that he has no means of crossing the water. He is not daunted; he can make the measurement by calculation.

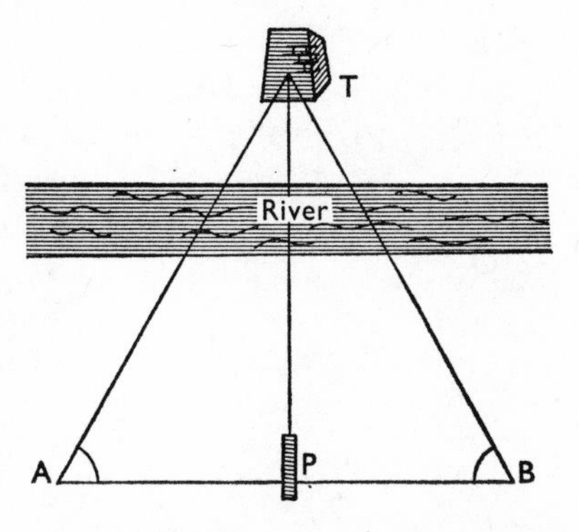

Fig. 15. Measuring the distance of an inaccessible object. The tower is T; the observer's reference point is P; the baseline AB.

First he marks out a baseline AB, with P in the middle of it and AP and TP at right angles to each other. Using his theodolite, he measures the angles TAB and TBA. Since the three angles of a triangle add up to 180 degrees, the third angle (ATB) of the big triangle can be found at once. Now we have found out all the angles of the triangle, and we also know the length of the baseline AB, because we can measure it directly. By drawing or by calculation we can now obtain all the distances in the triangle. The line TP splits the big triangle into halves, and so the length of TP can be found as well. The problem is complete, and we have the distance of our tower. The angle ATP (or BTP, which is equal to it) is known as the parallax of the point T with reference to the baseline.

If you want an even simpler example, hold up a finger, shut one eye, and then line up your finger with some relatively distant object, such as a chimney-pot. Now, without moving your head, use your other eye—and you will see that the lining-up of your finger with the chimney-pot is no longer exact, because you are looking from a slightly different direction. Going back to the diagram, your finger is represented by T, your nose by P, and your two eyes by A and B. If you could measure the apparent shift, you could find the parallax, and complete the calculation as before.

This is straightforward enough when the object to be measured is fairly close, but with greater distances you need a longer baseline, as otherwise the shift due to parallax will be too slight to be measured at all. Bradley knew this, and he realized that the Earth, with its diameter of less than 8,000 miles, would not be large enough. The best solution was to make use of our yearly motion round the Sun (Fig. 16).

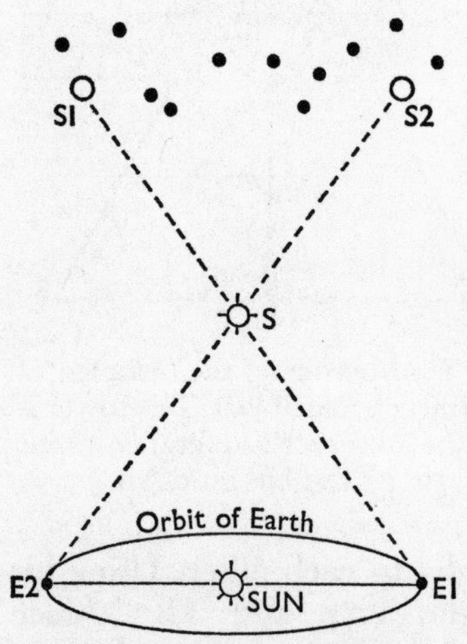

Fig. 16. Parallax as applied to the stars; S, the nearby star, is measured with reference to the background of more remote stars. For the sake of clarity, the drawing is shown very much out of scale. The actual parallax shifts of the stars are very small indeed—less than a second of arc.

The distance between the Earth and the Sun is, in round figures, 93 million miles, so that the diameter of the orbit is 186 million miles. If S is a comparatively near star, and we can regard the background stars as being infinitely remote, the parallax of S will be measurable. Observe it from alternate ends of the Earth's orbit (E1 and E2)—that is to say, at an interval of six months. From E1, the star will be seen at S1; from E2, at S2. We can measure the angles, just as our surveyor did, and work out the distance of the star.

This again is easy in theory, but unfortunately there are many complications. To begin with, the Sun itself is not fixed in space, and this introduces an error at the outset. We must be careful to select a close star, which is not easy when you do not know the distance of any of them. Moreover, the parallax angle is bound to be very small, and cannot amount to as much as one second of arc—the apparent length of a foot rule as seen from a distance of 39 miles. Bradley had set himself a tremendous problem, and, incidentally, the diagram given here is hopelessly out of scale. Not surprisingly, Bradley failed—but he did make a discovery which was as important as it was unexpected.

His target star was Gamma Draconis, which is of about the second magnitude. Bradley selected it because it passes directly overhead at Greenwich, and this meant that he could study it regularly by means of a special telescope fixed in a vertical position; he was in effect looking straight upward at the star as it crossed the overhead point or zenith.

Bradley found apparent movements indeed, but they did not seem to be due to parallax. He was badly puzzled, and ordered a new telescope which allowed him to examine other stars as well. All of them showed tiny shifts of the same kind. This in itself proved that parallax was not the cause, and then, one day when Bradley was out sailing on the Thames, he found the answer. If the direction of the boat were changed slightly, the mast-head shifted, not because of a change in the wind direction but because of an alteration in the boat's course. Many people have unconsciously noticed something of the sort. If you go out in a shower of rain, and the drops are falling almost vertically, you will have to tilt your umbrella forward if you want to avoid being drenched as you walk along. Only if you stand still will you need to hold the umbrella straight above your head.

Light does not travel instantaneously; it has a velocity of 186,000 miles per second. Therefore, the light coming from a star will always show an apparent displacement toward the direction in which the Earth is moving. Since the Earth's rate of motion round the Sun is 18½ miles per second, and the direction is changing all the time simply because our path is practically circular, the stars will show regular annual shifts, returning at the end of one year to their original positions. The effect is termed 'aberration'.

Bradley's discovery of aberration made him famous, and had a great deal to do with his appointment as Astronomer Royal when Halley died in 1742, but it did not help in solving the original problem, and when Bradley himself died, after twenty years in office, the distances of the stars remained as baffling as ever. By then a young Hanoverian musician had arrived in England—a musician who was to become perhaps the greatest observational astronomer of all time, and who in turn did his best to work out the scale of the universe.

Friedrich Wilhelm Herschel, always remembered as Sir William Herschel, began his career as a bandboy in the Hanoverian Guards. Military life did not appeal to him, and when he came to England he set out to make a career as an organist and music teacher. For some years he was organist at the Octagon Chapel in Bath (which, alas, no longer stands), but he developed a burning interest in astronomy, and over the years he built reflecting telescopes which were the best of their time. The largest of them had a mirror 49 inches in diameter. In 1781 he discovered a new planet, the one we now call Uranus, and was appointed King's Astronomer—not Astronomer Royal, by the way—so that he was able to give up music as a career and spend the rest of his life working for science.

Herschel was nothing if not methodical. Helped by his sister Caroline, he set out to "review the heavens", his main idea being to find out how the stars are distributed in space. Like Bradley, he proposed to measure distances by the parallax method, but his procedure was rather different, as he planned to make use of double stars.

Pairs of stars are common in the sky. Some of them are visible without a telescope; for instance if you look closely at Theta Tauri, close to the bright orange-red Aldebaran, you will see that it is made up of twins. Mizar or Zeta Ursæ Majoris, the second star in the tail of the Great Bear, has a fifth-magnitude star (Alcor) close beside it, and any small telescope will show that Mizar itself is made up of two components, one rather brighter than the other. Herschel's telescopic sky-sweeps yielded a rich harvest, and by 1785 he had published two catalogues of double stars, raising the total known number to over seven hundred. He was ready to begin his main task.

Let us suppose that two stars, A and B in Figure 17, lie in more or less the same direction in space as seen from Earth, but that B is much more remote than A. The effect will be that of a double star, as shown in position B_1. If the two components are equally luminous, B will naturally appear fainter than A, though since the stars have a very wide range in luminosity it does not always follow that a bright star is nearer than a dim one.

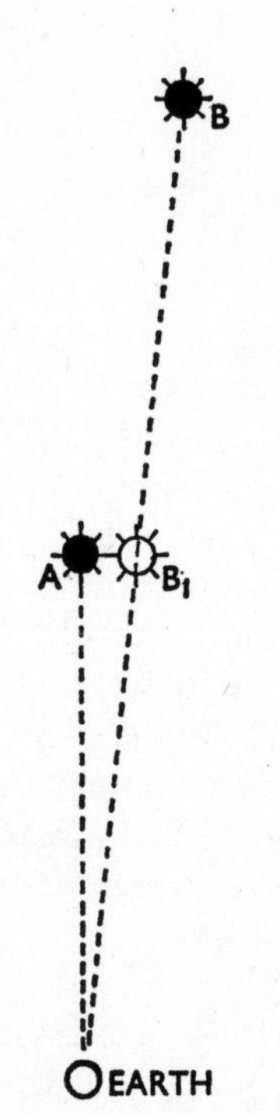

Fig. 17. An optical double. Stars A and B are not close together, but lie in the same direction as seen from Earth, and appear side by side in the sky.

On Herschel's reasoning, it should often happen that A would be close enough to show measurable parallax, while B would not. Therefore, B could be regarded as a stationary background marker, and the yearly shift of A could be measured relative to it. When solar motion, aberration and all other complications had been taken into account, the distance of A would be found.

It all seemed feasible, but no shifts came to light. The brighter components of double stars obstinately refused to show parallaxes as Herschel had hoped. This was indeed curious, and Herschel started again, re-measuring many of the pairs in his catalogues. Like Bradley half a century before, he made a discovery which was unexpected. In many cases the components of double stars appeared to be moving round each other, much as the two bells of a dumb-bell will do when twisted by their joining arm. This, of course, is the way in which the Earth and Moon move together round the Sun—though the Earth is so much more massive than the Moon that the 'balancing point', or centre of gravity of the combined system, lies inside the Earth's globe.

The inference was obvious. For such pairs, the components

were physically associated, and lay at the same distance from us. Herschel had discovered true star-pairs, or 'binaries'.

Of course, not all double stars are binaries. There are cases where the arrangement is much as shown in Figure 17, but these false or optical doubles are rather rare, while binary pairs are extremely common. Herschel had to admit defeat. We know now that his instruments, good though they were, were not accurate enough to measure stellar parallaxes.

Herschel died in 1822. During the following fifteen years the problem of stellar distances was taken up by three more observers, all of whom obtained positive results; they were Thomas Henderson in South Africa, Friedrich Bessel in Germany, and F. G. W. Struve in Estonia. The methods which they used were basically the same, though different in detail, and Bessel was the first to publish his results, though actually his measurements were made some time after Henderson's.

Bessel had already made his mark in the astronomical world. He was appointed Director of the Observatory of Königsberg at the early age of twenty-six, and he set himself to tackle problems of star-cataloguing. At that time Bradley's catalogue was the best in existence, and Bessel began to overhaul it, extending it at the same time, until at last he produced a list of the positions of 63,000 stars. Some of these stars showed proper motions, and in particular there was a quick-moving star of the fifth magnitude in the constellation of Cygnus, the Swan. Bayer had not thought it important enough to be given a Greek letter, and it was known as 61 Cygni, because this was the number which it had been allotted by Flamsteed. Its proper motion amounted to just over five seconds of arc per annum, which means that it would take over 350 years to shift by a distance equal to the apparent diameter of the Moon, but even so it was exceptionally rapid—and, presumably, exceptionally near. Moreover it was a binary, with the two components far enough apart to be separated with the aid of a small telescope.

Bessel began to study the star in 1837. Only a year later he was able to prove that both the components of 61 Cygni showed a parallax of 0″·3 (0·3 seconds of arc), corresponding to a distance of almost 11 light-years. Since the modern value is 10·7 light-years, Bessel's estimate was remarkably accurate.

Measured in our everyday units, 61 Cygni is roughly 60 million million miles away. Look at it in 1974, and you are really seeing it as it used to be in 1963.

One-third of a second of arc is a staggeringly small angle, but there are not many stars closer than 61 Cygni (a list is given in the Appendix). Of course, the greater the distance, the smaller the parallax shift. Most of the stars are too remote to yield any reliable shifts at all.

Henderson's selected star was Alpha Centauri, in the southern sky. The choice was a good one; Alpha Centauri's proper motion is comparable with that of 61 Cygni, and it also is a wide binary, though instead of being obscure it shines as the brightest star in the whole sky apart from Sirius and Canopus. It cannot be seen from Europe, but Henderson was Director of the Cape Observatory, and in 1832 he began his measurements. Unfortunately his health was not good, and he was compelled to retire from the Directorship after a comparatively brief spell in office; he came back to his native Scotland, and did not work out his results for Alpha Centauri until after the publication of Bessel's work for 61 Cygni. Actually Henderson had the easier task, since Alpha Centauri has a parallax of 0″·76, and is the closest of the bright stars, with a distance of 4·3 light-years or roughly 25 million million miles. We now know that Proxima, a very faint member of the Alpha Centauri system, is one-tenth of a light-year closer still, and is our nearest stellar neighbour with the obvious exception of the Sun.

Struve, at Dorpat in Estonia, concentrated upon the lovely bluish star Vega, in Lyra. Here the parallax is smaller, because Vega is further away, and Struve's results were less accurate; we now know that Vega is 26 light-years or over 150 million million miles from us. Yet even this is not far on the cosmical scale. To take a few other examples among bright stars; Capella has a parallax of 0″·073 (corresponding to 45 light-years), Regulus in Leo 0″·039 (85 light-years), and Spica in Virgo 0″·021 (220 light-years).

Light-years are convenient, but I must here introduce another unit, which is always quoted in technical papers. This is the parsec, or distance at which a star would yield a *par*allax of one *sec*ond of arc. It is equivalent to 3·26 light-years, or more than 19 million million miles. This is 206,000 times the so-called astronomical unit, or distance between the Earth and the Sun.

The work by Bessel and his contemporaries was of tremendous importance. The scale of the star-system had been found, and the first parallax measurements were soon followed by others. Yet the method has its limitations; for distances over 150 light-years the parallax shifts are too small to be measured really accurately, and by 600 light-years they are swamped by unavoidable errors in observation, so that the system breaks down.

To carry our distance-gauging further into space, we must therefore make use of less direct methods. By itself, the telescope cannot help us; we have to combine it with other instruments, the most important of which is the spectroscope. This brings us on to what we call astrophysics, or the physics of the stars, but before going into more detail it seems only right to give a brief description of the various constellation patterns. The stars become so much more interesting when you can learn to tell which is which.

Chapter Five

THE CONSTELLATIONS

THE BEST METHOD of learning your way around the night sky is to equip yourself with a star-map, go outdoors, and pick out the groups one by one. It is not nearly so difficult as it may sound, and once you have recognized a few 'skymarks' the rest will follow. I remember that when I set out to learn the stars (at the age of about eight) I made a pious resolution to identify one new group every night; and the method worked rather well, so that I am quite prepared to recommend it!

Though the constellations themselves do not change, you will sometimes see a starlike object which is not on your map. This will certainly be a planet, and may cause temporary confusion, since it will alter the whole look of that part of the sky; but all bright planets keep to the Zodiac, which is a great help in identifying them.

Moreover, each planet has a 'personality' of its own. Venus shines either in the western sky after sunset or in the eastern sky before dawn, and is so much brighter than anything else that it cannot be mistaken; at its best it may even cast a shadow. Jupiter, also, far outshines any star, while Mars is distinguished by its strong red colour, and Saturn by its steady, rather yellowish light. It is true that Saturn, and Mars when at its faintest, can easily be mistaken for stars, but any almanac will tell you where they are, and the more remote planets are never conspicuous, so that for the moment we need not trouble about them.

By no means all the constellations contain any objects of note, and even some of the Zodiacal groups are decidedly dull. Moreover, observers who live in the northern hemisphere are deprived of some of the finest constellations in the whole sky—which leads us on to some mathematical reckoning which, I am glad to say, does not involve anything more abstruse than ordinary subtraction.

Remember that declinations are reckoned according to the celestial equator, so that—for instance—Mizar in Ursa Major has a value of +55°, or 55 degrees north, while Sirius is at −17°, or 17 degrees south (I am, of course, giving round numbers). If you know your own latitude on the Earth, you can easily work out how much of the sky can be seen; all you have to do is to subtract your latitude from 90 degrees. Lizard Point, at the southernmost tip of England, has a latitude of +50°. Taking 50 away from 90, we are left with 40; therefore any star north of declination +40° will be circumpolar, while any star south of declination −40° will never rise at all. Since the brightest star in the Southern Cross has a declination of −63°, it is hopelessly and permanently out of view from anywhere in England—or, for that matter, anywhere in Europe. Vega, at +39°, is not quite circumpolar from the Lizard, and sets for a very brief period each day, whereas Capella, at +46°, is visible all the time, though when at its lowest a very slight horizon haze will conceal it.

Now let us take a trip to the Shetland Isles, where the lattitude is +60°. Here our limiting declination will be −30° (since 60 subtracted from 90 equals 30), and stars north of declination +30° will be circumpolar; Vega as well as Capella will remain permanently above the horizon. On the other hand, consider Fomalhaut in Piscis Austrinus (the Southern Fish), with its declination of almost −30°. It was easily visible from Lizard Point, but from the Shetlands you will be very lucky to see it.

Suppose we want to see the lovely southern star Canopus, which is second only to Sirius? We can make our calculation the other way round. The declination of Canopus is −53°; taking 53 away from 90 we are left with 37, so that we must go down to latitude 37° north. Gibraltar (latitude 35° north) will not do, but from Cairo (30° north) Canopus will reach a peak altitude of 7 degrees above the horizon, so that at its best it will be clearly seen—assuming, of course, that the observer is well away from any city lights.

These calculations can be made for any latitude, and in Appendix II I have given the declinations of the brightest stars in case anyone wants to do some checking. Actually, southern observers have the best of matters. Round the south celestial pole there are a great many bright stars and interesting objects, and it may be true to say that people who have seen the whole sky are usually ready to exchange our Bears, Dragon and Lynx for the Southern Cross, the Ship and the Centaur.

For a start, let us consider the sky as it appears from England and the northern part of the United States. I do not propose to go into any great detail about star recognition, because I have done so elsewhere.* All I propose to do here is to point out the main groups; anyone who is really interested can take the matter further.

(1) *NORTHERN HEMISPHERE*

Stars of the Far North (Fig. 18.)

Ursa Major, marked by its seven Plough or Big Dipper stars, is circumpolar from Britain (or the northern States), and is so easy to recognize that it makes a splendid 'signpost'. All the Plough stars have commonly-used proper names as well as their Greek letters; in fact the star at the end of the tail has two names—Alkaid and Benetnasch—as well as its official designation of Eta Ursæ Majoris. Megrez is fainter than its companions, while Mizar is the famous binary. Alcor, close to Mizar itself, is very easy to see with the naked eye except when conditions are ruined by cloud, mist or artificial lights.

The diagram shows some of the constellations round the north celestial pole which can be found by using the Bear. The Pointers, Merak and Dubhe, lead us to Ursa Minor (the Little Bear) which looks a little like a twisted and anæmic version of the Great Bear, but is celebrated because it contains Polaris, the Pole Star. The magnitude of Polaris is almost exactly 2. Beta Ursæ Minoris or Kocab, nicknamed the Guardian of the Pole, is about the same brightness. One point should be noted; under moonlight or misty conditions, the fainter stars of the Little Bear will not show up. Kocab, by the way, is decidedly orange in colour.

Next, we can find Cassiopeia. In mythology she was the proud queen of the Perseus legend; in the sky she takes the form of a kind of W or M of stars, of which the brightest are of the second magnitude. An easy way to find Cassiopeia is to pass an imaginary line from Mizar through Polaris, and continue it for about the same distance on the far side. The Milky Way

* In my book *The Amateur Astronomer* (Lutterworth, 1974), in which I have given detailed star-maps together with lists of interesting objects.

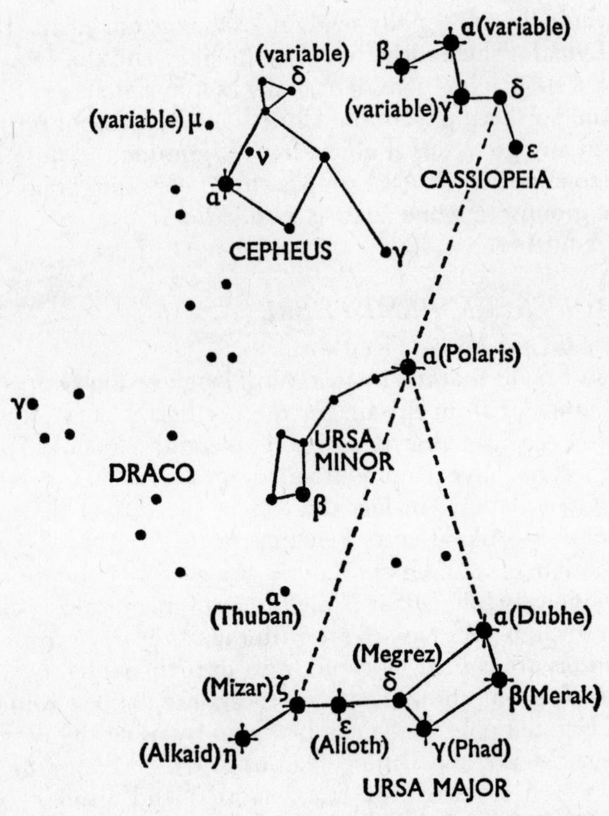

Fig. 18. Ursa Major, Ursa Minor, Cassiopeia and Cepheus.

flows through Cassiopeia, so that the whole region is rich in faint stars. Cepheus is much less prominent than his wife, but is notable because the group contains a star which has been of great importance in astronomy; it has no special name, and is known only by its official designation of Delta Cephei. It is variable, but never brighter than Megrez, the faintest member of the Plough, so that it does not stand out. Finally, I must mention the long, sprawling group of Draco (the Dragon); Alpha Draconis, or Thuban, was the Pole Star of ancient times, but is below the third magnitude.

Spring Evenings: say mid-April at about 9 p.m. (Fig. 19.)

The Great Bear is almost overhead. Following through the

'tail' we come to Arcturus in Boötes (the Herdsman), which can hardly be missed, since with the exception of Sirius it is the brightest star ever visible in England; it is a lovely light orange in colour. The rest of Boötes is unremarkable, but it is worth looking for Corona Borealis (the Northern Crown), a semi-circlet of stars which really does look rather like a crown. How anyone could make a herdsman out of Boötes is something which I cannot explain, but the ancients were nothing if not imaginative.

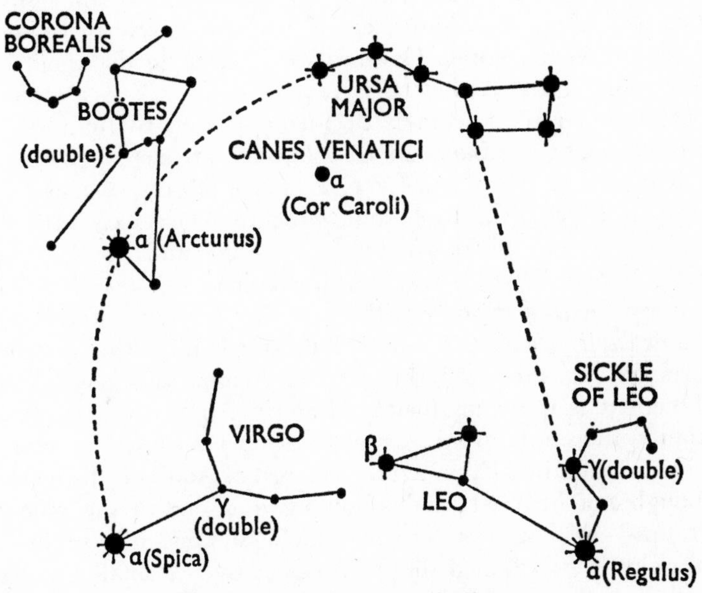

Fig. 19. Boötes, Leo and Virgo.

Following the curve still further we come to Virgo (the (Virgin), where the leading star, Spica, is of the first magnitude. Virgo takes the form of a distorted Y; note Arich or Gamma Virginis, which is a spectacular binary. The components are exactly equal, so that here we have a case of true stellar twinning. Not far off is Leo, the Lion; Regulus (magnitude 1) is the chief star, and the Sickle, shaped rather like a reversed question-mark, is distinctive. Gamma Leonis is a fine double. Between Leo and Ursa Major lies the third-magnitude Cor Caroli, in the small constellation of Canes Venatici (the

Hunting Dogs); here too is Coma Berenices (Berenice's Hair), which has no bright star, but is very rich in faint stars and remote galaxies. To the naked eye it looks almost like a large star-cluster.

During spring evenings Cassiopeia is at its lowest in the north; Capella lies in the west and Vega in the east. Orion, which dominates the evening sky during winter, has set.

Summer Evenings: say mid-July at 11 p.m. (Fig. 20.)

The Great Bear lies in the west, with Arcturus still visible; Leo and Virgo have gone, and Capella is barely to be seen, very low in the north. Overhead lies Vega, in the small but interesting constellation of Lyra (the Lyre); it is almost as bright as Arcturus, and is decidedly bluish, so that it is a glorious sight in binoculars or a low-power telescope. Also in this group we have Beta Lyræ, a binary of very special type; Epsilon Lyræ, a multiple star, obscure when seen with the naked eye but telescopically fascinating; and the so-called Ring Nebula, which I will describe later.

Vega forms a triangle with two other first-magnitude stars, Altair and Deneb; I once nicknamed this 'the Summer Triangle', and the term seems to have come into general use, though it is quite unofficial. Altair in Aquila (the Eagle) is high in the south, and is easy to recognize because it is flanked to either side by a fainter star. The rest of Aquila is distinctive, though not brilliant, and adjoining it is one of the modern groups—Scutum, the Shield, which contains a magnificent star-cluster nicknamed the Wild Duck. Even a small telescope will show that it contains a great many 'birds'.

Deneb, not so conspicuous as Altair though in fact much more luminous and remote, leads Cygnus, the Swan. Cygnus would be more aptly termed the Northern Cross, as indeed it often is; the X-shape is rather spoiled by the fact that one member, Beta Cygni or Albireo, is too faint and too far from the centre, but to make up for this Albireo proves to be probably the loveliest double star in the sky. The primary is golden-yellow, the companion bluish-green. I never tire of looking at it through my telescope. Of the smaller constellations in this area, take special note of Delphinus (the Dolphin), which is compact enough to be prominent even though it is not brilliant;

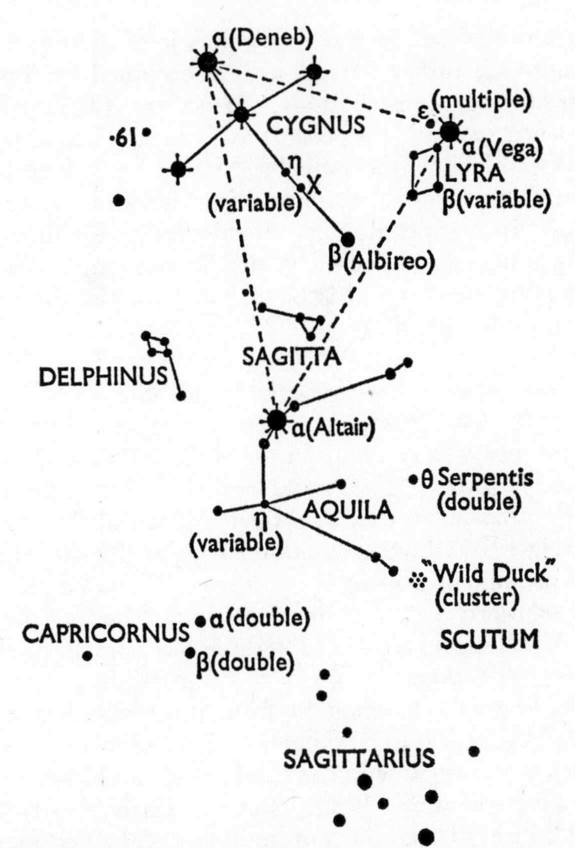

Fig. 20. The 'Summer Triangle'. Cygnus, Lyra and Aquila.

it was here that the English amateur George Alcock, in 1967, discovered a remarkably interesting new star or nova about which I will say more in Chapter 11.

Summer is a good time for looking at the Milky Way, which flows through Cassiopeia, Cygnus and Aquila down to the southern horizon. Low in the south you can see two large, bright constellations which are never at their best from Europe. Sagittarius (the Archer) has no really distinctive shape, but Scorpio or Scorpius (the Scorpion) is truly magnificent; its brightest star is Antares, of the first magnitude and strongly red in hue. Unfortunately Scorpio is so far south that part of it never rises at all over Britain.

The area enclosed by imaginary lines joining Arcturus, Vega and Antares is rather barren, as it is occupied by three large and sprawling constellations: Hercules, Ophiuchus (the Serpent-bearer) and Serpens (the Serpent). Rasalhague, or Alpha Ophiuchi, is the only bright star here, but there are some notable telescopic objects in Hercules, while Theta Serpentis is a fine double. Incidentally, Ophiuchus and Serpens appear to be locked in deadly combat. Judging from the old maps the Serpent is having the worst of the fray, and has been pulled in half.

Autumn Evenings: say mid-October at 9 p.m. (Fig. 21.)

This is the least spectacular time of the year. Ursa Major is low in the north, Vega high in the west, and Capella easterly; Arcturus, Regulus, Antares and Spica are below the horizon, and Orion has not yet risen, though Aldebaran in Taurus can be seen. Deneb (which is circumpolar from Britain) and Altair remain prominent.

The southern aspect is dominated by Pegasus, the Flying Horse. Needless to say it looks nothing like a horse, airborne or otherwise, and takes the form of a square; it is not so conspicuous as might be thought from the maps, because most newcomers expect it to be smaller and brighter than it really is. Well below it, very low in the south, you should be able to see Fomalhaut, the only bright star in Piscis Austrinus (the Southern Fish). This is the most southerly of the first-magnitude stars visible in Britain, and it barely rises in North Scotland, though from England it can be quite prominent.

Cassiopeia is very high up, and it is worth noting that two of the W-stars show the way to Pegasus. Leading off from Pegasus is the line of stars marking Andromeda, the princess of the famous legend; the main stars here are of the second magnitude, and Andromeda also contains the spiral galaxy M.31, which is just visible to the naked eye if you know where to look for it. Alpha Andromedæ, or Alpheratz, is actually one of the four members of the Square of Pegasus. It used to be known as Delta Pegasi, but for some reason or other was given a free transfer.

Beyond Andromeda, in the direction of Capella, we come to Perseus, the dashing hero of the legend. The main star,

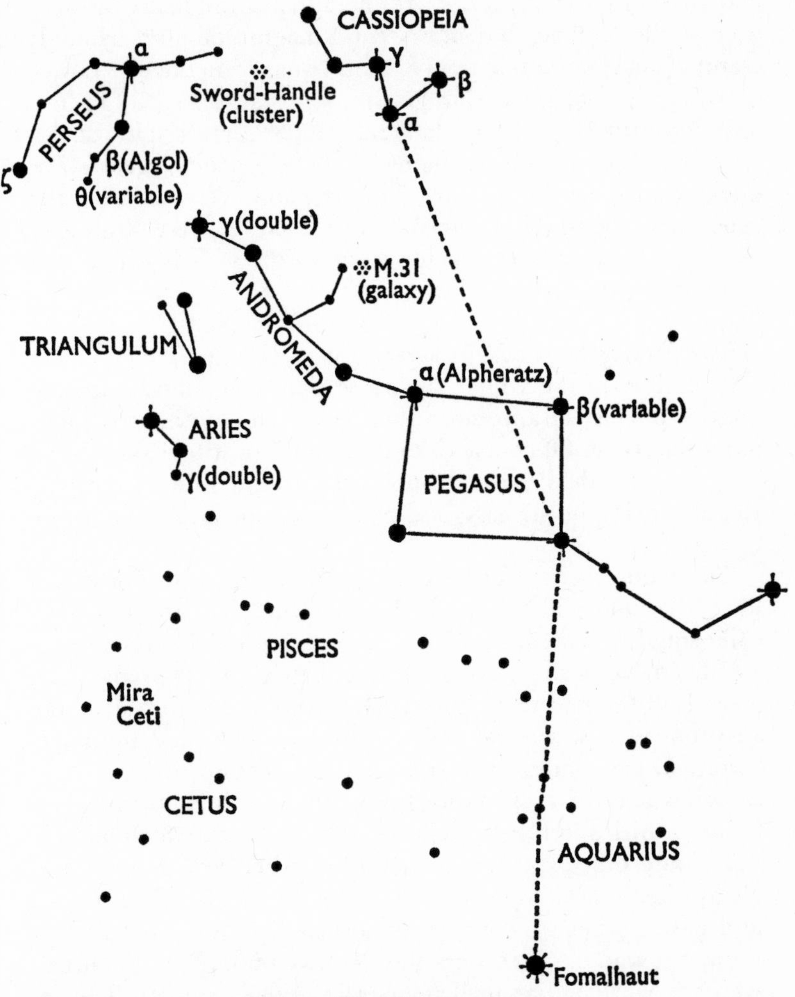

Fig. 21. The Pegasus area.

Mirphak, is only of the second magnitude, but the constellation as a whole is quite easy to locate, particularly since it is involved in the Milky Way. Two objects should be singled out. The so-called Sword-Handle is a superb double cluster; and Algol, or Beta Persei, is the prototype 'eclipsing binary', about which more anon.

Below Andromeda you can find Triangulum (the Triangle)

and Aries (the Ram). Aries, still counted as the first constellation of the Zodiac, has one second-magnitude star, Hamal; Gamma Arietis is a fine double with equal components. Lower down are several large, dim groups, such as Pisces (the Fishes) and Aquarius (the Water-bearer). Cetus, the Whale—sometimes regarded as the sea-monster of the Perseus legend—has one second-magnitude star, Diphda, and also includes the celebrated long-period variable Mira. At its best Mira can outshine the Pole Star, but for much of the year it is too faint to be seen with the naked eye.

Winter Evenings: say mid-January at 9 p.m. (Fig. 22.)

The glory of the winter sky more than compensates for the relative paucity of autumn. Orion lies in the south, and dominates the scene. Of course our old friends are still visible; Vega low in the north, Pegasus dropping toward the western horizon, Regulus rising in the east, and so on—but all these pale before Orion.

The Hunter's pattern cannot be mistaken. Of his individual stars, the pure white Rigel is the most brilliant, and is practically equal to Capella and Vega, though it is much more remote and far more powerful. By contrast, Betelgeux* is orange-red; it too is of the first magnitude, and is somewhat variable. Note also the three white stars of the Hunter's Belt, and the misty Sword which contains the great gaseous nebula.

Orion acts as a splendid guide. In one direction his Belt points to Sirius in Canis Major (the Great Dog), which far outshines any other star; it is white, but when low down—as it always is, as seen from Britain—it seems to twinkle strongly, and flash various colours. In the other direction, Orion's Belt shows the way to Aldebaran, the red 'Eye of the Bull' (Taurus), which is very similar to Betelgeux in appearance. In Taurus, too, are two naked-eye clusters: the Hyades, round Aldebaran, and the Pleiades or Seven Sisters. The Pleiades, particularly, form a glorious group, best seen with binoculars.

* This name may be spelled in several ways; Betelgeuse and Betelgeuze are other variants. Moreover, nobody seems to know quite how to pronounce it. I refuse to call it "Beetlejuice", as some people do; the name comes from the Arabic, and Arab scholars tell me that "Bay-tell-jurz" is about as near as we can get. According to R. H. Allen's classic work *Star-Names and their Meanings*, the original form was "Ibt al Jauzah", or the Armpit of the Central One.

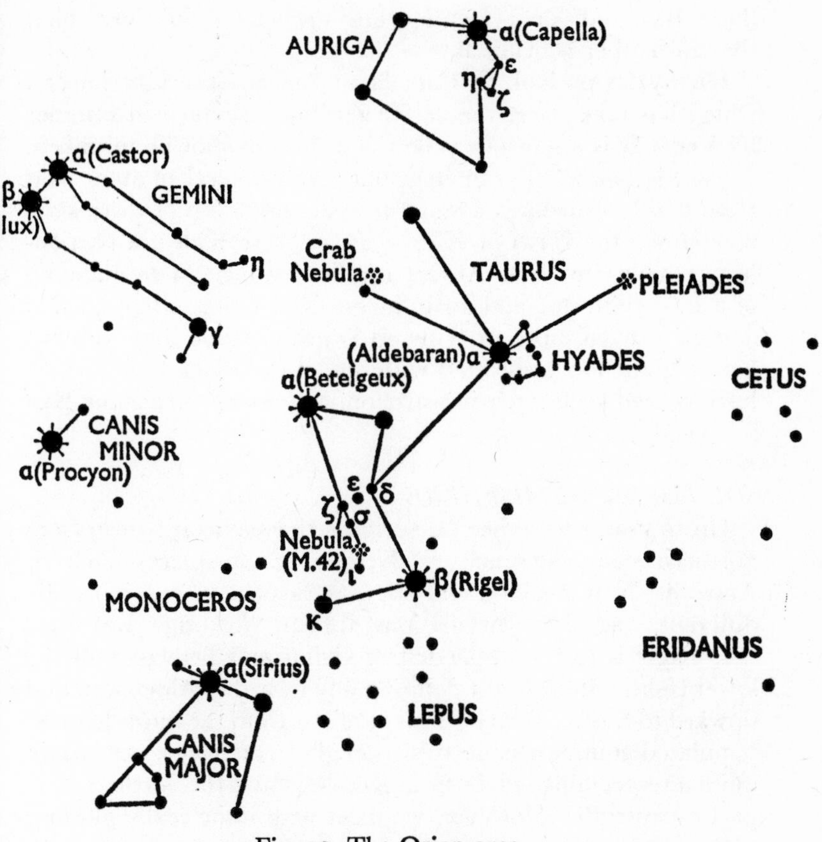

Fig. 22. The Orion area.

Orion's junior Dog, Canis Minor, is marked by the first-magnitude Procyon, and not far off are Castor and Pollux, the Heavenly Twins who have given their names to the constellation Gemini itself. Here also we have a legend. It is said that of the two brothers, Pollux was immortal, while Castor was not. When the inevitable happened, and Castor was killed, Pollux was so grief-stricken that the Olympians came to the rescue, and placed both youths in the sky. Castor, the fainter of the two, is white—perhaps as a result of his misfortunes!—while Pollux is orange-yellow.

The rest of Gemini is made up of lines of stars stretching from Castor and Pollux in the general direction of Betelgeux.

Both Eta and Zeta Geminorum are interesting variables, though neither is brilliant.

Finally let us look at Capella in Auriga (the Charioteer), which has taken over the zenith position occupied in summer by Vega. It is a glorious yellow star, and cannot be mistaken. Auriga is marked by an irregular quadrilateral of stars, and close beside Capella is a small triangle made up of three stars nicknamed the Hædi or Kids. Two of these Kids are particularly noteworthy. Zeta Aurigæ is a fascinating system made up of a hot white star and an immense Red Giant, while Epsilon Aurigæ is even more extreme. It is just possible that we have here a giant star associated with one of the Black Holes which have caused such a stir in astronomical circles during the past few years.

SOUTHERN HEMISPHERE

Up to now I have been describing the sky as it is seen from northern countries such as Europe. From places such as Australia, New Zealand and South Africa the view is naturally different, and the constellations appear 'the other way up'; thus Rigel is to the upper left of Orion and Betelgeux to the lower right, with the Belt pointing downward to Aldebaran and upward to Sirius. The two Bears are lost from the most densely-populated southern countries, though Ursa Major can attain quite a respectable height from Rhodesia and the northernmost part of Australia. However, we meet with some really glorious groups, such as Centaurus, Argo, and of course Crux Australis, the Southern Cross.

Most of the charts given earlier in this book can also be used by southern observers, though they have to be inverted. I do not propose to give a chart of the south polar region itself, because there is nothing of interest there; it is a very barren region. The south pole itself lies in the totally formless and obscure constellation of Octans, the Octant, and there is no bright star anywhere near it.

Summer Evenings: say mid-January at 9 p.m. (Fig. 23.)

Remember that the southern seasons are reversed from those of Europe, with Christmas coming in the hottest part of the year. Orion is high up, almost due north; Capella is visible above the northern horizon, though it can never be seen to advantage. Sirius is magnificent, and much more striking than it ever appears from Europe, because it is higher up. Almost at the overhead point lies Canopus, leader of the old constellation of Argo, the Ship; since Argo has been split up, Canopus has been reclassified as Alpha Carinæ. It is extremely luminous, with a surface rather cooler than that of Sirius. It is said to be yellowish, though I admit that to me it always looks white.

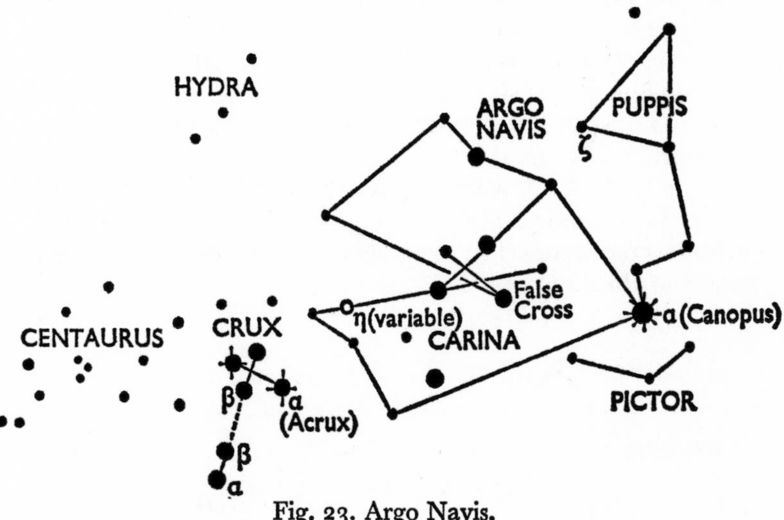

Fig. 23. Argo Navis.

Carina, the Ship's Keel, contains many other bright stars, though none to rival Canopus; and here too we have Eta Carinæ, the most erratic variable star in the sky, which was very brilliant in the mid-nineteenth century, though for many years now it has been invisible with the naked eye. The rest of the Ship—Vela (the Sails) and Puppis (the Poop)—is also rich, though without any really distinctive pattern.

Also high up is Achernar, of the first magnitude, which is the leader of the long, sprawling constellation of Eridanus, the River. Between Achernar and the barren south polar region can be seen two patches which look superficially like broken-off parts of the Milky Way. These are the famous Clouds of Magellan, which are separate galaxies, and about which I will have more to say later.

The Southern Cross is visible in the south-east, together with part of the Centaur, but winter is not the best time of the year for appreciating it.

Autumn Evenings: say mid-April at 9 p.m. (Fig. 24.)

Orion is setting in the west, though Sirius and Canopus are still quite high. In the north Leo has come into view, with Regulus now at the top of the Sickle instead of the bottom; Virgo can also be seen, and Scorpio is rising in the east. But the chief glory of autumn skies is Crux, together with Centaurus. Crux may be the smallest of all the constellations, but it contains four bright stars close together: two of the first magnitude, one just below the first magnitude, and one just below the third. To be candid, Crux is more like a kite than a cross, but it is none the less striking for that. It is immersed in the Milky Way, and contains a variety of interesting objects, including the wonderful 'Jewel Box' cluster and a dark nebula known as the Coal Sack.

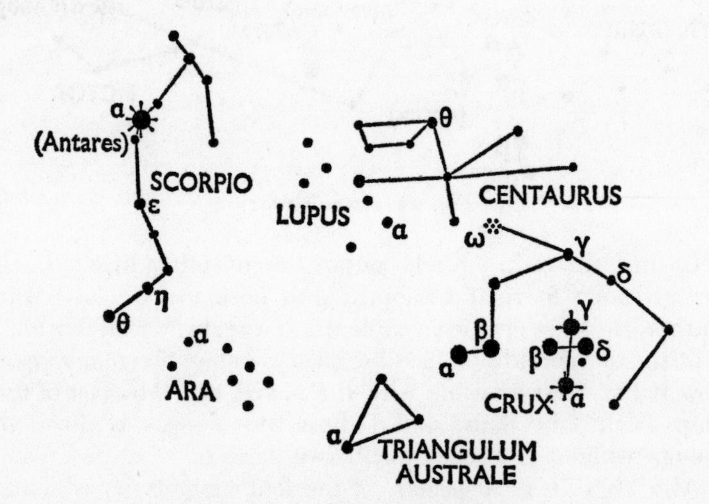

Fig. 24. Centaurus and Crux.

Centaurus, which more or less surrounds Crux, is a splendid constellation in its own right. It contains two stars of the first magnitude, of which the senior—Alpha Centauri—is the nearest of our stellar neighbours. (Proxima is known officially as Alpha Centauri C.) Beta, or Agena, is also very brilliant, and the two make excellent pointers to the Southern Cross. Also in the constellation are many other bright stars, as well as Omega Centauri, the finest globular cluster in the sky; and not far from the Pointers lies Triangulum Australe, the Southern Triangle, with the reddish second-magnitude Alpha.

Achernar is still to be seen, rather low in the south-west. Very high up the huge, faint constellation of Hydra, the Watersnake, sprawls across the sky; adjoining it the little quadrilateral of Corvus, the Crow, appears much more conspicuous than it ever seems from Europe.

Winter Evenings: say mid-June at 9 p.m.

Orion is now completely out of sight, and Sirius too has gone, though Canopus remains above the horizon. Crux and Centaurus are high; so too is Scorpio, which is truly magnificent with its long stream of stars and the brilliant red Antares. Following it round is Sagittarius (the Archer), containing the superb star-clouds which indicate the direction of the centre of the Galaxy. Arcturus is visible in the north, with Virgo still to be seen in the west. From Rhodesia, the Great Bear can be found low in the north, though from South Africa and most of Australia it is never visible. The two Clouds of Magellan may be seen between the south pole and the horizon; Hercules, Ophiuchus and Serpens are rising in the north-east.

Spring Evenings: say mid-September at 9 p.m. (Fig. 25.)

This is the best time of the year to see Vega, Altair and Deneb, though only Altair climbs to a respectable altitude. The Square of Pegasus is rising in the north-east, Scorpio sinking in the west; Canopus is at its lowest, almost down to the southern horizon, while Crux and Centurus too are descending.

Overhead you can see the rather distinctive form of Grus, the Crane, which gives some impression of a flying bird; of its two main stars, both ranking of the second magnitude, Alpha (Alnair) is white, while Beta is warm orange. Nearby are several other constellations named after birds: Tucana (the Toucan), Pavo (the Peacock) and Phœnix (the Phœnix), but there are few bright stars or notable objects apart from the superb globular cluster in Tucana.

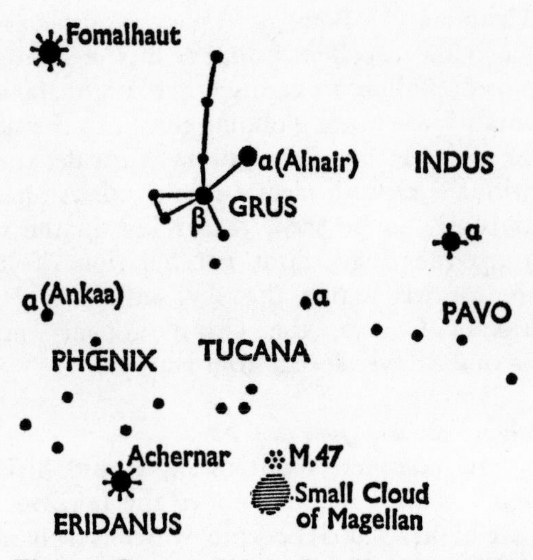

Fig. 25. Grus and the other 'Southern Birds'.

Fomalhaut in the Southern Fish is high and is very prominent, partly because it is bright and partly because it is relatively isolated. Northern-hemisphere observers are often surprised to see how conspicuous it becomes when really well placed.

I know very well that this review of the sky is very sketchy and incomplete, but at least I hope that it will serve as a basis. If you make up your mind to learn your way around, you should have little difficulty; and whether you live in London or Sydney, New York or Auckland, Oslo or Tierra del Fuego, you will find that there is always plenty to interest you among the stars.

Chapter Six

THE MESSAGE OF STARLIGHT

IN 1825 A FRENCHMAN, August Comte, wrote a book called *Cours de Philosophie Positive* in which he made one particularly profound statement. Some things, he said, are destined to remain permanently unknown to mankind; and as an excellent example he cited the chemistry of the stars. According to Comte, it was absolutely impossible to find out "what stars are made of".

Other statements of like nature have been shown to be equally wrong. In 1840 Dr. Dionysius Lardner, addressing the British Association, gave his opinion that "men might as well try to reach the Moon as to cross the stormy North Atlantic Ocean by means of steam power"—and let us not forget that even after the Wright brothers had made their first short-range hops above the ground, a leading American astronomer, Simon Newcomb, proved to his own satisfaction that flying in a heavier-than-air machine was totally out of the question. But going back to 1840 we find François Arago, one of the most famous astronomers of the day, repeating the theory supported by Sir William Herschel, according to which the Sun has a cool region beneath its hot surface, and that intelligent beings could well live there.

Some people are credulous enough to believe almost anything (who has not been tackled by flying saucer enthusiasts, for instance?) but the fact that Herschel and Arago could accept such ideas is a pointer to the paucity of our knowledge less than two centuries ago. Since then we have learned a great deal, and in fact the first steps in studying the chemistry of the stars had been taken by Sir Isaac Newton more than a hundred and fifty years before Comte wrote his book.

In 1666 Newton, then young and unknown, was living at his home at Woolsthorpe, in Lincolnshire; he had temporarily left Cambridge because of the Great Plague, which had resulted in the very wise decision to close the University until the danger was over. At Woolsthorpe, Newton busied himself in laying the foundations for much of his later work. Some of his main studies concerned the nature of light.

One reason for this interest was that he was anxious to build better telescopes. Up to then, all telescopes had been of the refracting type, and were not satisfactory because of the false colour problem. Newton determined to find out just why this false colour appeared.

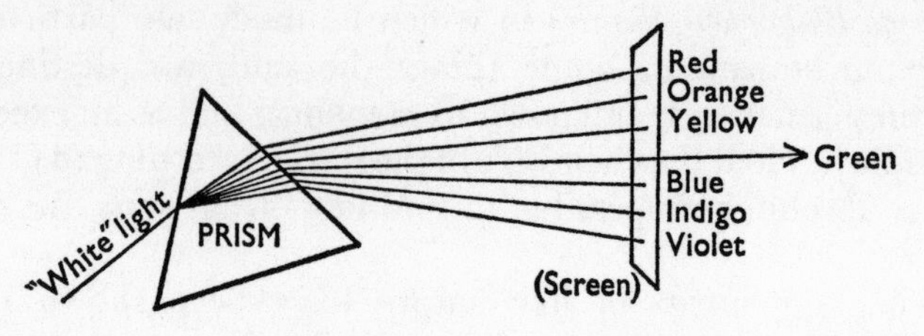

Fig. 26. Production of a spectrum. The prism splits up the ray of light into its constituent colours, from red at the long-wave end of the band through to violet at the short-wave end.

What he did was to make a hole in an opaque blind, and admit a beam of sunlight, which he then passed through a glass prism. When it emerged from the prism, the light was spread out into a rainbow, from red at one end of the band to violet at the other (Fig. 26.) Newton then placed a second screen with a hole to admit the light of one colour only, and passed this one colour through another prism. This time there was no rainbow. The ray was slightly bent or 'refracted', but it remained the same colour as before.

This gave Newton the key to the whole problem. Sunlight, like all so-called 'white' light, is really a mixture of all the colours of the rainbow, and the glass prism splits it up. The violet part of the mixture is refracted more sharply than the blue, the blue more than the green, and so on until we reach red, which is refracted least of all. Consequently, the different colours are spread out to give a luminous band or 'spectrum'. In the case of a single colour, of course, no such effect will be seen, since we are no longer dealing with a mixture.

This explained the cause of the false colour which had so puzzled the early workers. An *object-glass* in a refractor acts rather in the same way as a prism, inasmuch as it bends the different parts of the mixture unequally; the blue light is brought to focus closer to the object-glass than in the case of the red (Fig. 27). Newton saw no way round the difficulty. He therefore abandoned refractors altogether, and built the first telescope of the reflecting type. Here there was no false colour, since a mirror reflects all parts of the mixture equally.

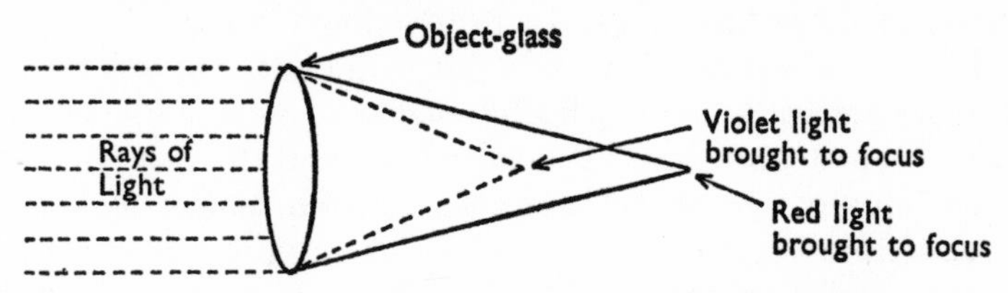

Fig. 27. Production of false colour by an object-glass. The violet part of the light is brought to focus closer to the object-glass than the red. (Not to scale.)

This was as far as Newton went. He did not realize that object-glasses can be improved by making them compound, and neither did he follow up his studies of the Sun's spectrum. Even his theory of light was criticized, and some of the correspondence of that period is not without its humorous side. In Volume 10 of the *Philosophical Transactions* of the Royal Society, for instance, we find the following entry: "A Letter of Mr. Franc. Linus, written to the Publishers from Liège on the 25th of Febr. 1675, being a Reply in the letter printed in Numb. 110, by way of Answer to a former letter of the same Mr. Linus, concerning Mr. Isaac Newton's Theory of Light and Colours." On the next page appears "Mr. Isaac Newton's Considerations on the former Reply" . . . and so on. Tempers on both sides were apt to become frayed, and one has to admit that the scientists of the seventeenth century were no less quarrelsome than those of the twentieth.

Little more work on the solar spectrum was done for many years, but in 1802 an English physicist, W. H. Wollaston, repeated Newton's experiment, using a slit to admit the sunlight. He found seven dark lines crossing the coloured band, but he thought that these lines simply marked the boundaries between the various hues—and in consequence missed the chance of making a great discovery. This honour came to a young German named Josef von Fraunhofer.

Fraunhofer was born at Straubing, in Bavaria. Both his parents died when he was still very young, and his schooling was fragmentary. At the age of fourteen he was apprenticed to a Munich looking-glass maker, one Weichselberger. We often hear tales about cruel taskmasters and starving, ill-treated apprentice boys, but in Fraunhofer's case the description fitted the facts. Then, one day, the tumbledown house in which he lodged collapsed in a heap of rubble; the accident was seen by the Elector of Bavaria, who happened to be driving by, and the Elector took it into his head to befriend the boy, who had been injured. He gave Fraunhofer enough money to buy his release from Weichselberger, and to take up the study of optics.

Fraunhofer's ability soon showed itself. In 1806 he obtained a post at the Optical and Physical Institute at Munich, and his reputation spread. He constructed a special instrument known as a heliometer, used by Bessel to measure the distance of 61 Cygni years later, and he became Director of the Institute in 1823, though unfortunately he died three years later at the early age of forty. He was the maker of a 9½-inch object-glass, the largest of its time, which was bought by the Russian Government and installed at Dorpat in Estonia; the telescope was clock-driven, another development which was revolutionary. F. G. W. Struve's main work, including his measurement of the parallax of Vega, was carried out at Dorpat.

About 1814 Fraunhofer began the research for which he is now best remembered. Like Wollaston, he attached a spectroscope to the eye-end of a telescope, and re-observed the mysterious dark lines in the solar rainbow. His instruments were so much better than Wollaston's that instead of seeing only seven lines, he could make out several hundreds; evidently they did not merely mark the boundaries between different colours, but were far more significant.

Fraunhofer was deeply interested. He realized that the lines were fixed and constant; for instance, a prominent double line in the yellow part of the band was always present, and he wondered whether it might be associated with the element sodium, since luminous sodium vapour shows a *bright* double line, yellow in colour. Having charted 574 lines in the solar spectrum, he turned his attention to the stars, and obtained equally fascinating results. Here too the general effect was of a rainbow band crossed by dark lines, but in some cases the familiar solar lines were lacking, while new ones were seen in different positions along the band.

No doubt Fraunhofer would have carried on his work, but he was given no time to do so. For a quarter of a century after his death the dark lines remained unexplained. The problem was finally solved in 1859 by Gustav Kirchhoff, Professor of Physics at the German university at Heidelberg, who laid down the three fundamental laws of spectroscopy which bear his name. These laws are so important that I must say rather more about them.

The first law is straightforward enough. It states that incandescent solids or liquids, and also incandescent gases under high pressure, produce a 'continuous' spectrum—that is to say, a rainbow band.

The second law states that a luminous gas or vapour under low pressure will produce an entirely different effect. Instead of a continuous strip, there will be various isolated bright lines—and each line will be the trade-mark of some particular element or group of elements. This is termed an 'emission' spectrum.

Matter is made up of atoms, which combine into groups or molecules. There are only a limited number of types of atoms, known as the elements; hydrogen, oxygen, iron and tin are typical examples. All material is made up of these fundamental substances, and we may be sure that no elements remain to be discovered in the series. At one time it was believed that there were 92 elements, hydrogen being the lightest and uranium the heaviest; recently, others have been tacked on to the heavy end of the sequence (that is to say, beyond uranium), but there are no 'missing numbers'. Atom-groups are known as molecules. For instance, a molecule of water is made up of two hydrogen atoms combined with one atom of oxygen—hence the chemical formula H_2O—while a molecule of common salt is made up of one sodium atom together with one atom of the element chlorine.

Now let us go back to Kirchhoff's second law, and consider the famous double line which had so interested Fraunhofer. Luminous sodium vapour produces this pair of lines; no other element can do so—and in consequence, whenever we see the double yellow line, we know that sodium must be responsible. It is the copyright of sodium, and sodium alone.

Each element produces a whole series of lines. Some elements are more prolific than others, iron, particularly, yielding many hundreds of lines; but since no two lines exactly coincide, it is theoretically possible to disentangle one element from another.

The heart of the dark-line problem is Kirchhoff's third law, and the best way to explain it is to picture a simple experiment. If you burn salt in a flame, you will produce sodium vapour, which will of course yield an emission spectrum containing the double yellow line.* If you look at the spectrum of an electric light bulb, you will find a continuous band, since the filament of the bulb is an incandescent solid (Law 1). Now take the bulb and put it behind the flame, so that you are looking at the emission spectrum of the sodium against the background of the continuous spectrum produced by the bulb. Instead of a rainbow with bright sodium lines superimposed upon it, what you will see is a rainbow crossed by *dark* lines. In fact, the atoms in the sodium vapour are removing part of the corresponding portion of the continuous spectrum, which is why the dark lines are known as Absorption Lines.

The crux of the matter is that the positions of the lines are unaffected, and this in itself means that they can be tracked down to the elements responsible for them. As soon as you remove the background bulb, the sodium lines in the flame become brilliant once more. Incidentally, the absorption lines are never truly black; they emit a good deal of light, and seem dark only because the rainbow background is brighter.

Such were Kirchhoff's Laws. Now let us apply them to the Sun.

The principle is exactly the same. In the background we have our 'bulb'—that is to say, the Sun's bright surface, which consists of high-pressure gas and yields a rainbow spectrum. In front we have the 'flame', represented by the shell of luminous gases above the solar surface. This shell, or chromosphere, contains incandescent sodium, and so the double yellow line, together with all the other lines in the sodium spectrum, appears dark (Fig. 28). Here is positive proof that there is sodium in the Sun's chromosphere. Comte was wrong after all!

* There will be many other lines as well, since common salt contains chlorine as well as sodium—not to mention various impurities; but to simplify matters, let us consider only the double yellow line due to sodium.

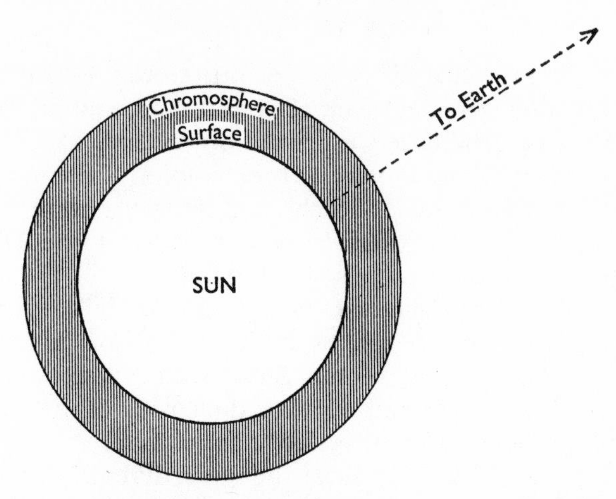

Fig. 28. Production of the Sun's spectrum. The bright surface yields a continuous rainbow band; the chromosphere yields a line spectrum, and as seen from Earth these lines are reversed, so that they are seen as dark (absorption) lines rather than bright (emission) lines.

Nowadays around seventy elements have been identified in the Sun's spectrum. One story is worth quoting here, since it gives extra proof that astronomy, even in the nineteenth century, was far from being a useless and abstract science. In 1869 the English astronomer Norman Lockyer was examining the spectrum of the Sun when he noted that one line in the orange-yellow region did not correspond with any element known at the time. He suggested that it might be due to an unfamiliar element, and proposed to name it helium, from the Greek word for 'sun'. A quarter of a century later another Briton, Ramsay, discovered helium on the Earth; it proved to be the lightest of all the elements apart from hydrogen. (Unlike hydrogen, it is not inflammable, which is why it was once used for filling the gasbags of airships.)

The Sun is much the nearest of the stars, and so its spectrum can be studied in great detail. Normally the astronomer's cry is for "More light!" and for this reason great telescopes are built to collect as much light as possible; but with the Sun there is no such difficulty. Stellar spectra present problems of a very different order, since even with the brightest stars there is not enough light for the spectrum to be really widely spread out. Fraunhofer, of course, made a start; he was followed by men such as Angelo Secchi, a Jesuit priest who examined the spectra of 4,000 stars between 1864 and 1868, and Sir William Huggins, who established a private observatory at Tulse Hill and concentrated upon very precise studies of the spectra of certain individual stars. Secchi and Huggins were in turn followed by E. C. Pickering, who worked at the Harvard College Observatory in the United States.

Gradually, some concrete facts emerged. The spectra of different stars were found to be by no means alike; some closely resembled that of the Sun, while others were very different. This was bound up with the colour of the star concerned, and hence with its surface temperature.

As I have already stressed, the stars differ in hue. Thus Betelgeux, Antares and Aldebaran are reddish, Arcturus orange, Capella yellow, Rigel and Sirius white, and Vega bluish. With the naked eye, the colours of the fainter stars cannot be seen—remember the old saying that "in moonlight, all cats look grey!"—but binoculars or telescopes bring them out well. For instance, an irregular variable star known as Mu Cephei was described by Herschel as 'garnet', and other observers have compared it with a glowing coal. Since white heat is greater than red heat, it is natural to believe that white or bluish stars are hotter than the orange or reddish stars.

Secchi divided the stars into four spectral classes. His nomenclature formed a useful basis, but since it is now obsolete there is no point in saying much about it except that Type I was made up of white stars, II of yellow or orange, and III and IV of red.

In 1890 Pickering, at Harvard, introduced a more detailed system, modifications of which have stood the test of time. The general idea was to divide the stars into spectral groups, lettering them A, B, C, D, and so on—starting with white stars and working through yellow, orange and orange-red to red. Inevitably the letters soon became out of order; Types C, D and E proved to be unnecessary, and the final result was alphabetically chaotic, so that the modern 'spectrum alphabet' runs: W, O, B, A, F, G, K, M, R, N, S. The mnemonic "Wow! Oh Be A Fine Girl Kiss Me Right Now Sweetie" (or Smack) is well known; a certain amount of mild amusement may be gained from deriving others!

At least the series is logical in one way, since it makes up a true sequence. In the white A-type stars, for instance, the spectral lines due to hydrogen are very prominent; they are less intense in the next type (F), fainter still in G, and very inconspicuous in K. This does not necessarily mean that K-type stars contain less hydrogen than those of type A, but merely that conditions are not so suitable for the hydrogen to show itself—though, let me add at once, the stars differ widely in chemical composition. Some, for example, are very poor in metals.

Each type is divided up into sub-classes, usually numbered from 0 to 9. To take part of this order at random, beginning with (say) A0, we have A1, A2, A3 . . . A9, F0 and so on. A star which is midway in type between A0 and F0 will therefore be classed as A5; there is very little difference between B9 and A0, or between A9 and F0.

A vast amount of work has gone into this classification, and we must admire the work of those early pioneers who had to carry out all their observing visually. Today, of course, photography has taken over.

I do not propose to go into more detail about the refinements of the spectral classification, but it seems worth while to give a few notes about the main types—beginning with the hottest stars, which should logically have been given the letter A, but are in fact allotted W.

W stars. These are almost in the nature of celestial freaks, since they show rainbow backgrounds crossed by many lines which are bright instead of dark. These emission lines are due mainly to helium, carbon, nitrogen and oxygen. The surface temperatures are vey high, of the order of 80,000 degrees C; all W stars are very luminous and remote. They are known as Wolf-Rayet stars, after the two astronomers who first described them in detail. Allied to them are the *O stars*, with surface temperatures of around 35,000 degrees, and with both bright and dark lines in their spectra. Both these types are rather uncommon. Gamma Velorum (O7) and Zeta Orionis (O9) are among the brightest examples.

B stars. Bluish white (B0) to white (B9); temperatures from 25,000 degrees (B0) to 12,000 degrees (B9). There are no emission lines, but absorption lines of hydrogen and (particularly) helium are dominant. Examples: Epsilon Orionis (B0), Rigel (B8).

A stars. Known commonly as Sirian stars, since Sirius is of this type (A1). Temperatures from 10,000 to 8,000 degrees. A-type stars are white, and their spectra are dominated by lines due to hydrogen. Other examples are Vega (A0) and Altair (A7).

F stars. Yellowish, with surface temperatures from 7,500 to 6,000 degrees. Hydrogen is less prominent in the spectra than with type A, but calcium is conspicuous; note two calcium lines lettered H and K. Typical F-type stars are Procyon (F5) and Polaris (F8). The brilliant Canopus also has an F-type spectrum, but is not a completely ordinary star, as it is exceptionally luminous.

G-stars. Yellow; the Sun is of this type (G2). Hydrogen lines continue to weaken, but lines due to metals are numerous and prominent. With this type we find a definite division into very luminous 'giants' and more modest 'dwarfs'; the two classes are distinct, and do not merge into each other. I will have much more to say about this giant-and-dwarf division below. Typical stars are Capella (giant, G8) and the Sun (which, as we have already noted is ranked as a dwarf). Temperatures: 5,500 to 4,200 degrees for giants, and 6,000 to 5,000 degrees for dwarfs.

K-stars. Orange. Since Arcturus belongs to this type, the whole class is sometimes nicknamed 'Arcturian'. Metallic lines are strong, and those of hydrogen are very weak; the H and K lines due to calcium are still in evidence, and the temperatures range from 4,000 to 3,000 degrees for giants, 5,000 to 4,000 degrees for dwarfs. K-stars are among the most numerous of all. Typical examples are Arcturus (K2), Pollux (K0) and Aldebaran (K5).

M-stars. Orange-red; temperatures about 3,000 degrees for giants and 3,200 degrees for dwarfs. The spectra are very complicated, with many bands due to molecules. There is a tremendous difference between the giants and the dwarfs. Typical examples: Antares (M1, giant), Betelgeux (M2, giant), and Proxima Centauri (M5, dwarf). Many M-type giants are variable in light.

R, N and S stars. All these are red and remote, so that they appear dim in our skies. The reddest of all are the 'carbon' stars of type N.

Listed in this way, the differences between the various spectral types may seem obvious enough, but even a casual glance at a photograph of a stellar spectrum will show that interpretation is by no means easy. Remember, too, that not all stars can be put into neat, compact classes. For instance, the extraordinary White Dwarfs—about which I will have much to say later—show almost nothing apart from a few very broad lines; we have sub-giants, sub-dwarfs and so on. The 'variations on a theme' seem almost endless.

Yet an amazing amount of information has been drawn from studies made with the help of the spectroscope. We can find out how the stars are moving, and we can also estimate distances. From the spectra, it often happens that the luminosity ratio between two stars may be found, and this gives an immediate clue. Suppose that we have two stars A and B, A being close enough to show measurable parallax while B is not, and suppose that we also know how much more (or less) luminous B is than A. Our knowledge of A's distance then leads us on to that of B.

More important still, we have a means of probing into the real natures of these other suns. The message of starlight has come to make real sense to us.

Chapter Seven

A STAR'S SURFACE AND SURROUNDINGS

WE CAN MEASURE the temperatures of the stars. The different colours give a first clue—it is clear that the bluish Vega must be hotter than the yellow Capella, or the orange Betelgeux—but we can use the spectroscope to provide really accurate values.

Yet we cannot look closely on to a star's surface to see what is happening there. No disks are visible, and there is a limit to what we can find out simply by staring at a twinkling point. Fortunately this does not apply to the Sun, and if we are to make serious investigations with respect to the stars it is only sensible to begin near home. Let us, then, see what solar enthusiasts have to say.

The crux of the whole matter is that the Sun is perfectly normal, so far as we know. There is nothing to mark it out from millions of other stars in the Galaxy, and we have every reason to think that its surface is typical of many others. Instead of a tiny dot, we are presented with a dazzling disk, and there is as much light as any astronomer could want. In fact, there is almost too much, and to study the Sun we do not need huge telescopes.

It would be too much of a digression to go into details of the equipment used in observatories which specialize in solar research, and I will say only that the spectroscope, with instruments based upon the same principle, is all-important. But for the moment, consider only the features which are visible upon the Sun's disk. What you must never do, under any circumstances, is to point a telescope (or binoculars) at the Sun and then look through the eyepiece. You will concentrate all the heat on to your eye, and permanent blindness will result. I have given this warning more times than I can count—in books, on the radio and on television—and I make no apology for repeating it here. If you want to look at the Sun, use your telescope as a projector, and throw the Sun's image on to a screen held behind the eyepiece. Should you be given a dark sun-cap, with instructions to screw the cap over the telescope eyepiece for direct viewing, I can only recommend you to take the dark cap and drop it in the nearest pond.

With projection, however, there is no danger at all, and when the Sun appears on your improvised screen the disk will seem to be sharply-bounded. The bright surface, or photosphere, is at a temperature of 6,000 degrees, as we have already noted; but it will often be seen that there are dark patches on it here and there. These sunspots, first observed telescopically by Galileo and his contemporaries over three and a half centuries ago, are not so black as they look. Their surface temperatures are of the order of 4,000 degrees, and they seem dark only because they are cooler than the surrounding photosphere.

Sunspots are interesting to watch, and they are of immense significance in solar research; incidentally, they are associated with strong magnetic fields. Unfortunately it is out of the question to see 'starspots' on other suns, though I have no doubt that they exist. On the Sun itself, the whole photosphere is in a constant state of turbulence, and photographs show a mottled appearance called granulation—due to rising columns of gas from the still hotter layers below. On average, each granule is about a thousand miles in diameter, and the upward motions are over half a mile per second. The Sun's surface is never calm, despite its outward appearance.

The bright surface is overlaid by a layer of more tenuous gas, the chromosphere, which produces the dark Fraunhofer lines. Beyond the chromosphere we come to the corona, which is made up of excessively rarefied gas and is very extensive. It is not easy to study, even with special equipment, but luckily nature can come to our help. During a total solar eclipse the corona can be seen with the naked eye, because the Moon acts as a temporary screen.

The theory of an eclipse is easy to understand. By pure coincidence, the Sun and Moon appear virtually the same size in the sky. When the Moon passes directly between the Sun and the Earth (Fig. 29), it blots out the brilliant photosphere; and as soon as the glare has been cut off, the chromosphere and the corona shine out gloriously. We can also see the prominences, once known by the misleading name of Red Flames, which are clouds of incandescent hydrogen lying above the photosphere (Plate I). Special equipment makes it easy to study the prominences and the chromosphere at any time, but the outer corona is a very different problem, so that eclipses are still regarded as tremendously important. The trouble is that they are not as common as might be desired; the eclipse has to be 100 per cent total. The last to be visible from England took place in 1927, while the next will not be until August 11th, 1999. The reason is that the Moon's shadow only just touches the Earth, as shown in the diagram, so that the observer has to be in the right place at the right time; moreover, no total eclipse can last for as long as eight minutes, and most are far shorter. I have been fortunate enough to see four total eclipses, the last of which involved travelling to the coast of North Africa on June 30th, 1973, and there is no doubt in my mind that totality is the most glorious sight in all nature.

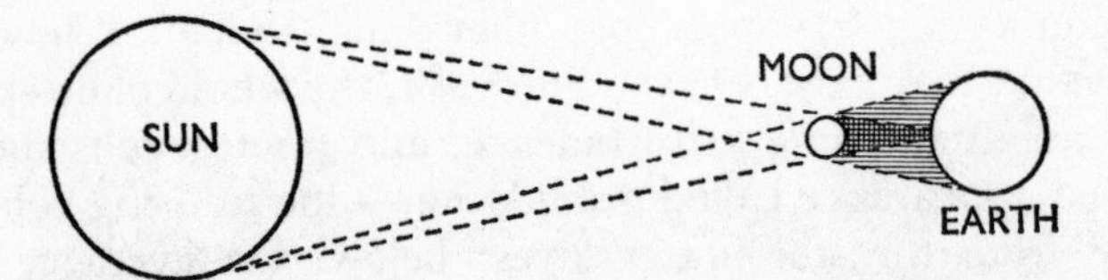

Fig. 29. Theory of a solar eclipse. The cone of the Moon's shadow just touches the Earth; to either side is an area of 'partial shadow'.

In the future, total eclipses will become less vital, because the Sun's surroundings can be studied at any time once you are above the top of the Earth's dense layer of atmosphere. A long programme of solar research was, for instance, carried out by the crews of the Skylab space-station in 1973. All in all, we have collected a tremendous amount of information, and we are in a good position to start comparing the Sun with other stars.

One term which is always cropping up is 'absolute magnitude'. This is the apparent magnitude that a star would have if taken out to a standard distance of 10 parsecs (32·6 light-years). Thus Rigel in Orion has an apparent magnitude of +0·1, while Sirius is brighter at −1·4; but Rigel is very remote, while Sirius is one of our closest stellar neighbours. From 10 parsecs, Rigel would have a magnitude of −7, and would cast strong shadows, while Sirius would be reduced to +0·7. Absolute magnitude, then, is a measure of actual luminosity. The Sun has an absolute magnitude of +4·8, so that from our standard distance it would be a very inconspicuous object as seen with the naked eye.

It may also be useful to mention colour index, which depends on the star's surface temperature. As we have seen, visual magnitude is a measure of a star's apparent brightness in ordinary light. Photographic magnitude is measured by the apparent size of the star's image upon a sensitive plate. Most early emulsions were blue-sensitive, and so a red star appeared comparatively faint. Subtract visual magnitude from photographic magnitude, and you have the colour index; thus if the visual magnitude of a star is 3·0 and the photographic magnitude is 5·0, the colour index will be 2·0—and the star will be red. There are any number of refinements, but the principle is clear enough. The scale is adjusted so that for a pure white star of spectral type A0, the visual and photographic magnitudes are the same, and the colour index is zero.

Next, I must introduce something which is of absolutely vital importance in all stellar astronomy: the Hertzsprung-Russell or H-R Diagram. It is, of course, bound up with the way in which a star produces its energy, but superficially it depends upon colour and luminosity.

It was due originally to two men: Ejnar Hertzsprung of Denmark (who died not so long ago at the advanced age of ninety-six) and H. N. Russell of the United States. In 1911 Hertzsprung made a plot in which he charted the stars according to their luminosities and their spectral types, and Russell elaborated it a couple of years later. From the typical H-R Diagram shown here, it is obvious that most of the stars lie in a diagonal band from the top left to the lower right; this band is always known as the Main Sequence. It begins with the very hot and luminous white or bluish stars; near the middle of the Sequence we find stars such as the Sun, and the dim Red Dwarfs lie at the lower end. Note also the so-called Giant Branch, to the upper right. But look for a red star of type M with about the same luminosity as the Sun, and you will not find it. Red stars come in two classes—very luminous or very dim, with none in between. The division between giants and dwarfs is less for type K, and less still for G and F, while for the 'earlier' types (A, B and O) it does not exist at all.

There is no chance that this sort of arrangement could be due to sheer coincidence. The giant and dwarf division is real enough, and it is of vital significance. I will say much more about it below, when we come to consider stellar evolution.

I have put in a few famous stars here and there over this particular diagram (Fig. 30); for instance Sirius—type A, luminosity 26 times that of the Sun; Spica, type B, luminosity 1500; Arcturus, type K, luminosity 100; Betelgeux, type M, luminosity 1200 (perhaps rather more, according to some estimates); Proxima Centauri, also type M, luminosity 0·0001;

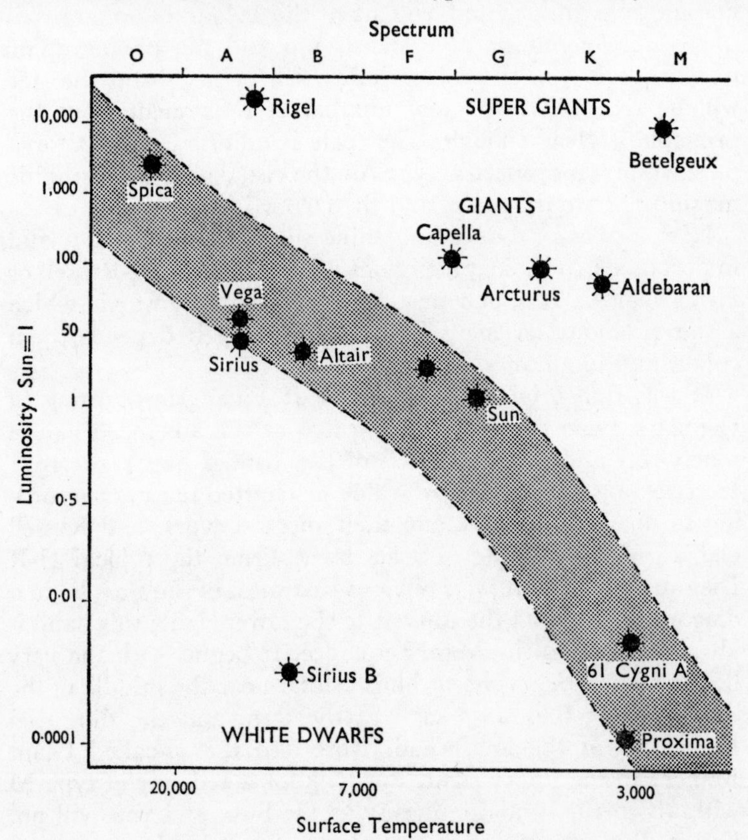

Fig. 30. The Hertzsprung-Russell or H-R Diagram, showing a few typical stars.

and the brighter component of the 61 Cygni pair, type K5, luminosity 0·06. The differences in diameter are as diverse as the luminosity ranges (Fig. 31). 61 Cygni has a diameter of

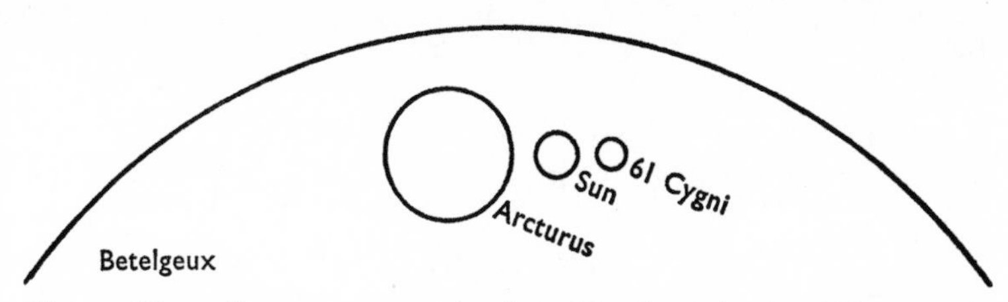

Fig. 31. Sizes of some stars; supergiant (Betelgeux), giant (Arcturus), Main Sequence red dwarf (61 Cygni)

600,000 miles, whereas Arcturus, also with a K-spectrum, is a full 26 million miles across. Some of the Red Giants are huge indeed; Betelgeux has a diameter of 250 million miles, so that it could more than swallow up the whole orbit of the Earth round the Sun. Yet the range in mass is much less, because the giants are always more rarefied than the dwarfs. Arcturus, for instance, is only sixteen times as massive as 61 Cygni.

Not all the stars fit snugly into the kind of H-R Diagram given here. In particular, the White Dwarfs, to the lower left, are very special cases. They are not ordinary white stars of low luminosity; they are well advanced in their life-stories, and are approaching death. The uncommon red stars of types R, N and S can, for our present purpose, be lumped together with type M. Then, too, we have the extremely hot Wolf-Rayet stars of type W.

The story of the W-stars goes back to 1867, when the first three were recognized by G. Wolf and G. Rayet at the Paris Observatory. The spectra were peculiar; the rainbow background was faint, and broad, bright bands dominated the scene. Other similar stars were soon found, including a few which are bright enough to be seen without a telescope—including the second-magnitude Gamma Velorum in the now dismembered Ship Argo, unfortunately too far south to be seen from Europe. There are two main classes of Wolf-Rayet stars. In the first, WN, there are bright spectral lines of hydrogen, helium and nitrogen, while the dark lines are broad and weak. The second class contains bright lines of hydrogen, helium, carbon and oxygen; because of the presence of carbon lines and the virtual absence of nitrogen, the type has been given the rather unhappy designation of WC.

All Wolf-Rayets are highly luminous—thousands of times brighter than the Sun—and all are remote. Their surface temperatures exceed 50,000 degrees, so that our Sun, with its 6,000-degree photosphere, seems very mild in comparison. Yet are we really justified in talking about the 'surface' of a Wolf-Rayet?

The dark lines in the spectra are shifted toward the violet or short-wave end of the rainbow band. This is certainly because of what we call the Doppler effect, and means that material is approaching us; in fact, it is streaming outward from the star at a velocity of over 300 miles per second. There is a constant loss of material, and it is hard to say just where the true surface ends and the extended atmosphere of the star begins.

I must digress for a moment in order to say something about the Doppler effect, because it too is so important. And as usual, I propose to give the familiar but hackneyed analogy of the whistling train—simply because I have never been able to think of anything better.

If you stand beside a railway line and listen to a whistling engine approaching you, the note of the whistle will be high-pitched. As soon as the train has passed by, and has begun to recede, the note of the whistle will drop. There is no mystery about it. During approach, rather more sound-waves per second are entering your ear than would be the case with a motionless train; the 'wavelength' therefore seems to be shortened. During recession, fewer sound-waves per second reach you, so that the wavelength seems to be lengthened and the note of the whistle falls.

It is much the same with light. The wavelength is shortened if the light-source is approaching, so that everything is a little too 'blue'; if the source of illumination is moving away, there is a slight reddening. The actual colour change is too slight to be noticed in everyday life (do not expect a red traffic-light to be seen as blue if you approach it at top speed!) but there is a perceptible effect upon spectral lines, as shown in the admittedly over-simplified diagram given here. In Fig. 32 (a), the position

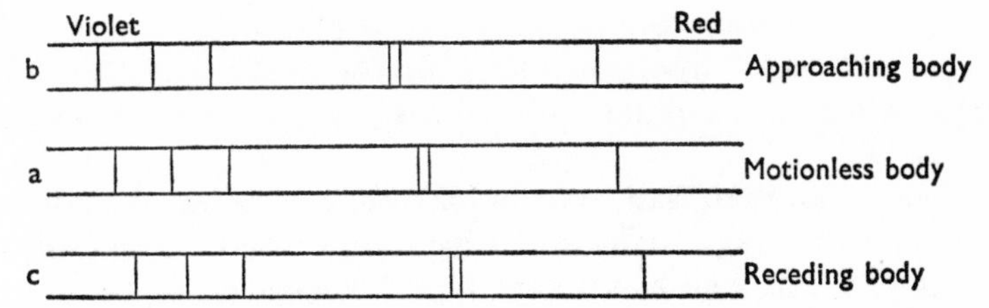

Fig. 32. The Doppler effect. The centre line shows the spectral lines of a motionless body. The upper line shows an approaching body, with the lines shifted to the violet; the lower line, a receding body, with the lines shifted to the red.

of some particular line is given on the assumption that the light-source is stationary relative to the observer. In (b) the source is approaching, and the line is displaced toward the short-wave or violet end of the spectrum; in (c) it is receding, and there is a Red Shift. The diagram is out of scale as well as being simplified, but it is enough to show the general principle.

This explains why the violet shift in the dark lines of a Wolf-Rayet spectrum shows that material is coming toward us from the star itself. We know, too, that an absorption line is produced by a hot gaseous layer seen against the background of the main body of the star. But what about the bright or emission lines? The only way to explain them is to suppose that much of the 'chromosphere' is *not* seen against such a background, and this confirms that the star's atmosphere is of tremendous extent. In such a

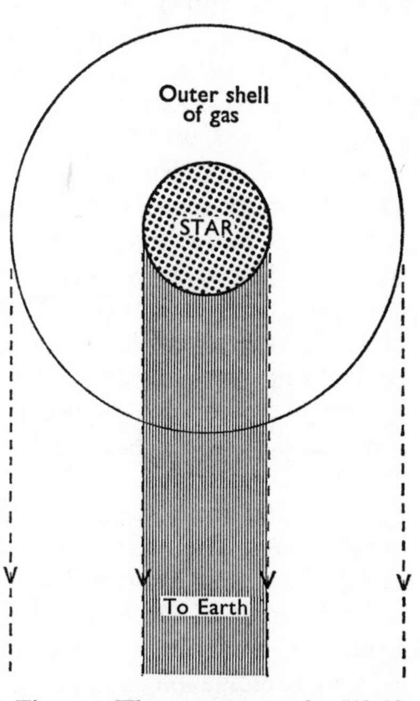

Fig. 33. The spectrum of a Wolf-Rayet star. The shaded area gives the usual kind of absorption spectrum; to either side, the gas is not seen against a stellar surface, and so produces emission lines.

case, the situation will be as shown in Figure 33; the shaded area will yield absorption lines in the usual way, but the remainder of the outer shell of gases will produce emission lines.

Note, too, that many W-stars are close binaries, the fainter component being usually a massive star of type O. The O-stars themselves, as well as some of type B (classified as Be) show some emission lines, also indicating the presence of an extensive shell of gas; in some cases where we find a binary system with one component of type W and the other of type O, it may be that there is a vast gaseous shell which includes both stars—truly a complicated arrangement, since it is even possible that there may be a flow from the 'atmosphere' of one star to that of the other. Yet true Wolf-Rayets are rare. About a hundred and fifty are known in our Galaxy, and another fifty in the southern external system known as the Large Magellanic Cloud. This means that, on average, less than one star in a million is a Wolf-Rayet, and we may assume that the state is a temporary one—which, in view of the constant loss of material, is not surprising. Nor even a very massive star can stay as a Wolf-Rayet for long on the cosmical time-scale.

We must admit that we do not yet understand as much as we would like to do with respect to the Wolf-Rayet stars, and there may well be a link with the so-called planetary nebulæ. Here we have a very hot central star, usually about as massive as the Sun, surrounded by a gaseous envelope which may have a diameter of up to 100,000 times the distance between the Earth and the Sun. Several hundred of these strange objects are known, of which the most famous is the Ring Nebula in Lyra, not far from Vega. Telescopically they appear as small disks or rings. Their names are inappropriate; they are not true nebulæ, and they are certainly not planets!

Another use of the Doppler principle is to show us whether a star is approaching or receding, and indeed we depend upon it for most of our information about the make-up of our Galaxy and others. We can also, in some cases, measure the rate at which a star is rotating on its axis. Unless we are looking straight at the pole of rotation, one limb of the star will be approaching us and will give a violet shift, while the other limb will be receding, so that the shift will be toward the red

(Fig. 34). The effect is easy to explain by means of a simple experiment. Take a cricket ball, and pencil in a spot and a cross separated by 180 degrees. If you now hold up the ball and spin it round, the spot will be approaching you when the cross is receding, and vice versa.

For our quickly-spinning star, then, the spectral lines will be displaced both toward the violet and toward the red; in other words, they will be spread out and broadened. If we see exceptionally broad lines, we may be fairly sure that the star is in rapid rotation. (It is, of course, obvious that all effects of this sort are superimposed upon the general Doppler shift due to the star's towards-or-away motion relative to the Earth.)

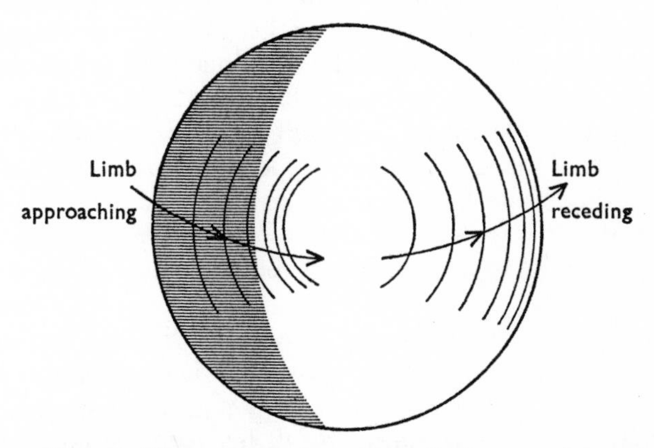

Fig. 34. When a star rotates, one limb must always be approaching us, showing a violet shift, while the other limb is receding from us, showing a red shift; the overall effect is to broaden the line—if, of course, we are looking more or less 'equator-on'. Looking 'pole-on', the effect would not be observed.

Our world is spinning, and so the equatorial zone bulges slightly, making the diameter as measured through the equator 26 miles greater than that as measured through the poles. The giant planet Jupiter takes only 9¾ hours to complete one turn, and the distortion is more marked; the equatorial diameter is 5,000 miles greater than the polar. With some stars, the rotation is so rapid that the star is drawn out into a shape which is not even approximately spherical.

All shell stars are quick-spinners. One of the most interesting, Pleione, may be taken as a good example. It is easy to identify, since it lies in the prominent cluster of the Pleiades or Seven Sisters; it has a B-type spectrum, and is of the fifth magnitude, so that it is visible with the naked eye. The rotation is about a hundred times faster than the Sun's, which is so rapid that in addition to the flattening there is a good chance that material will be flung off from the equatorial zone, forming a sort of gaseous ring—in shape not too unlike the ring-system around the planet Saturn, though of course quite different in character.

This may actually happen. One such major outburst occurred in 1938, and we have some idea of the sequence of events, though naturally the ring itself could not be seen. Some extra disturbance inside Pleione increased the acceleration of atoms outward from the star's photosphere, near the equator, and a temporary ring built up. This lasted for some years, but finally no more material became available to maintain the ring, and by 1952 it had vanished. Pleione had behaved in this way on more than one previous occasion, and we may assume that material is periodically hurled away from it.

We can also investigate the magnetic fields of the stars, because of yet another effect—named in honour of the man who discovered it in 1896: the Dutch physicist Zeeman. He found that when a light-source is put between the poles of an electromagnet, and the spectrum is examined, the lines will be split. In some cases there will be two components instead of the original line, one shifted to the red and the other to the violet of the original position; in other cases the main line will be accompanied by fainter components to either side. This is not the place to go into technicalities, but we can see that if a star has a powerful magnetic field its spectral lines will show the Zeeman effect.

The pioneer work was carried out by Horace Babcock, at Mount Wilson Observatory in California, in 1947. He concentrated upon stars which showed sharp absorption lines, so that presumably we were seeing their poles of rotation; quick-spinners with their equators presented to us would be affected by the conventional Doppler broadening of the lines. (I hope there is no confusion here between the Doppler and the Zeeman effects. To recapitulate: a Doppler refers to towards-or-away physical motion, a Zeeman purely to magnetism.) Also, Babock selected stars mainly of type A, because later-type stars are, in general, slower spinners. He was successful almost at once; to date 350 stars have been closely examined for these effects, and nearly a hundred have been found to have powerful magnetic fields. Some of these field-strengths are variable, though in general a 'magnetic variable' shows only slight optical fluctuations. The cause of the variations is not known with certainty. It may be that the magnetic axis is not the same as the rotational axis; or perhaps there is a definite cycle of activity. This is reasonable enough, because the famous 11-year cycle of the Sun is unquestionably associated with magnetic phenomena.

There is another solar phenomenon which is, apparently, closely linked with events upon at least some of the stars. A 'flare' is a temporary increase in brightness over a small area of the photosphere, usually associated with an active sunspot group; they are not often seen in ordinary light, but because they are made up of incandescent hydrogen they can be studied spectroscopically, and they are far from uncommon. They send out intense short-wave radiation as well as high-speed particles, and they are truly violent. For instance, one eruption associated with a flare was recorded on ciné film in 1951; it rose to a height of 30,000 miles at a rate of over 400 miles per second.

If we could look at the Sun from the distance of the nearest star, a flare would make no detectable difference to the total magnitude, but we now know of some stars where flares can more than double the output. The first of them, UV Ceti, was studied in 1948, and has given its name to the whole class. All flare stars are Red Dwarfs of type M, with very cool surfaces below 3,000 degrees. Because of their intrinsic faintness, we can see only those which are relatively close to us, though according to one estimate the Red Dwarfs make up 80 per cent of the total stellar population.

With a flare star, the visual magnitude may rise very suddenly, in only a minute or two, and then drop back to normal more slowly. UV Ceti itself once rose from its usual magnitude (13) to the fringe of naked-eye visibility with startling abruptness, remaining bright for only five minutes before starting to fade once more. Another flare star, admittedly rather less spectacular, is Proxima Centauri. Over twenty are now known. During outbursts, long-wavelength 'radio emissions' are also sent out.

There seems no reason to doubt that the sudden increases are due to flare activity on a gigantic scale. All the stars of the class are of very low mass, and have never been able to reach the stage of producing energy in the same way as the Sun. UV Ceti itself, like many of its kind, is a binary; each component has a mere 0·04 of the Sun's mass.

Even though a star appears as nothing more than a dot of light, we have found out a great deal about stellar surfaces and atmospheres. This, in turn, leads us on to the really fundamental problem of how a star evolves from birth to death.

Chapter Eight

THE LIFE OF A STAR

WHAT KEEPS THE Sun shining?

This is a question which has been asked time and time again, and all sorts of theories have been put forward. It is easy to say simply that the Sun is 'burning', but this is not the answer; for one thing, the Sun is too hot to burn! As we have seen, even at its surface the temperature is 6,000 degrees, and near the centre we can calculate that this temperature rises to the incredible value of about 14 million degrees.

The stars are suns; and like our Sun, they are sending out energy all the time. Naturally, their reserves are not inexhaustible, and so every star must have a definite life-cycle from birth to death. This means that they must evolve, changing their condition and their output as they do so. The trouble is that in general they change much too slowly for us to have the slightest hope of catching them in the act.

Let me give a homely comparison, and picture a visitor from (say) Mars, who has no previous idea of what an Earthman looks like. Having arrived here by some means or other, he takes a walk down Oxford Street in the rush hour. He will see infants, youths and men; he will not be able to watch a schoolboy turning into a man, but if he has any reasoning power he will soon be able to fit the various examples of *homo sapiens* into an evolutionary sequence, beginning with the babies and ending up with the greybeards. From this, he ought to be able to work out how a human being develops.

We are faced with much the same problem when we consider the stars. There are young stars, middle-aged stars and old stars; once we can arrange them properly, we will be on the right track. Unfortunately there is no easy way to tell which are the 'babies' and which are the 'senior citizens', and until 1939 or thereabouts most astronomers picked wrong. Red Giants such as Betelgeux used to be regarded as youthful, whereas now we hold the view that they are decidedly senile.

There is the further complication that not all stars go

through the same evolutionary sequence. Far from it! The story of a dim star such as, say, Proxima Centauri is quite unlike that of a huge globe of the Betelgeux type. Moreover, everything is much more complicated than used to be thought only a few decades ago.

Luckily, we have one or two established facts which we can use to build up a basis for our general picture, and of these the most valuable is our knowledge of the age of the Earth. There seems very little doubt that our world began its story between 4,500 and 5,000 million years ago, so that 4,700 million years is a reliable estimate; and the same applies to the Moon, as we have learned from analyses of the rocks brought home by the Apollo astronauts and the Russian Luna probes. This is a very long time indeed, as another condensed scale model will show. Suppose that we take the age of the Earth to be one week? On this scale, the dinosaurs were at their peak five or six hours ago; three hours ago, the tiny tree-living primates were ushering in the age of mankind; the Ice Age ended one and a half seconds ago, and it is only half a second since King Harold received an arrow in the eye during the Battle of Hastings. Whichever theory of the origin of the Solar System we adopt, it seems logical to assume that the Sun is rather older than the Earth, so that it has been in existence for more than 5,000 million years.

We also have reliable knowledge of the mass of the Sun, the temperature of the surface, and the rate at which energy is being poured out. And in consequence we can at once reject any idea that the Sun is burning in a conventional manner. Lord Kelvin, many decades ago, showed that a Sun made up of coal, and radiating as fiercely as the real Sun actually does, would burn completely away in only a few thousands of years. We must look for some process which can allow for a much longer time-scale.

It seems that a star begins its career by condensing out of an interstellar dust-and-gas cloud. We know of many such clouds, and term them galactic nebulæ, of which the best-known lies in the Sword of Orion. These nebulæ are stellar birthplaces.

The material in a nebula is amazingly tenuous—millions of times less dense than the air we breathe—but it cannot be quite uniform throughout its mass. There must be local condensations, and as each condensation becomes pronounced its own gravitational pull will come into play. Each particle will tend to attract its neighbours, and the result will be that the material will start to bunch together towards a definite centre, as well as drawing in fresh particles from the neighbouring regions.

Let us take up the story of a star whose mass is not very different from that of the Sun. Material is collecting in a definite zone inside the nebula, but it is not yet a true star, because it is not shining. As a starting-point, consider the time when our star has a mean radius of a thousand astronomical units—one astronomical unit, remember, being equal to the distance between the Earth and the Sun, or 93,000,000 miles. In approximately five thousand years (the time-scale is bound to be arbitrary) the radius has shrunk to only ten astronomical units, which is very roughly the same as the maximum distance between our Sun and the planet Saturn. Still the shrinking goes on, under the influence of gravity; when the radius has become six astronomical units, rather more than the mean distance of Jupiter from the Sun, there is a dramatic development. The star adjusts its structure, so that heat from the interior can be carried to the surface. In only a century or so, which is brief indeed on the cosmical scale, the star becomes from 100 to 1,000 times as luminous as our Sun is today; it has begun its main career. Convection is the main cause of this sudden change. The best comparison I can give (though not a really accurate one) is that of heated water beginning to circulate as its temperature rises.

Up to this stage the star has drawn its heat solely from gravitational contraction. It used to be thought that this process would be enough to sustain a star throughout its career, which led Sir Norman Lockyer, a leading English astronomer of the late nineteenth century, to put forward a theory of stellar evolution which sounded delightfully straightforward. Actually, Lockyer was wrong; nevertheless, his work paved the way for a better understanding of the stars.

Lockyer began by describing the way in which a star would condense out of a nebula (Fig. 35). The original temperature would be low, and the youthful star would be large, with a cool, red surface; Betelgeux and Antares would be good examples. As the contraction went on, the temperature would

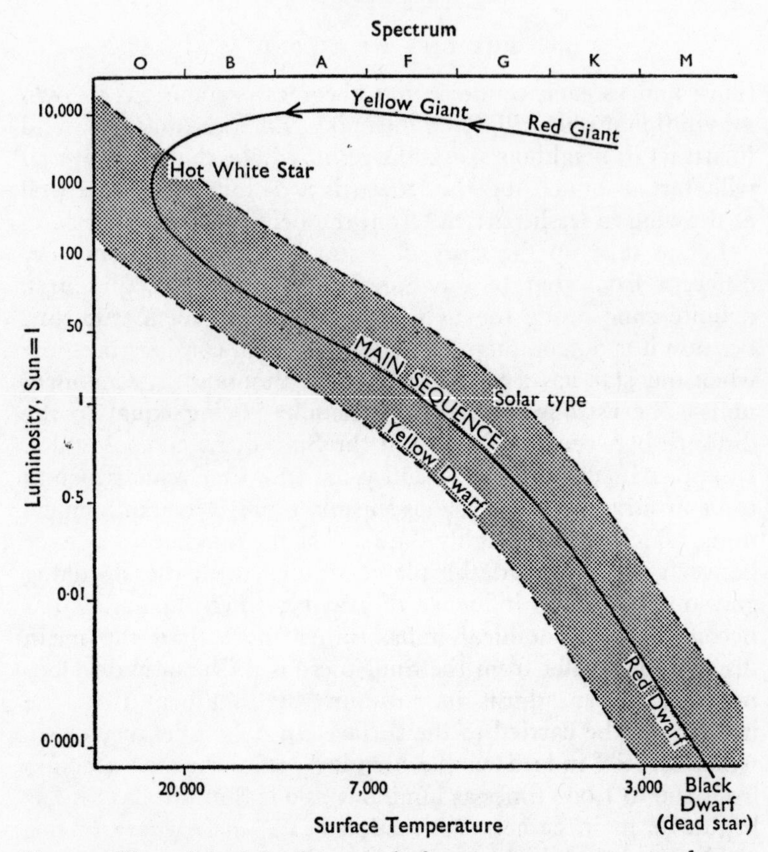

Fig. 35. The original Lockyer evolutionary sequence; a star began as a Red Giant, passed down the Main Sequence and ended up as a dead Black Dwarf. The theory seemed plausible, but it is now known to be completely wrong.

continue to rise; the star would become successively an orange giant (Arcturus), a yellow giant (Capella) and then a much smaller, highly luminous white star (Rigel). This would be the peak of the star's career. The gravitational contraction would continue, but the temperature would fall as well, and the star would pass steadily down the Main Sequence to become a less energetic white star (Sirius), a yellow dwarf (the Sun) and finally a red dwarf (Proxima). In extreme old age all its light and heat would leave it, and it would turn into a cold, dark globe.

Lockyer's theory accounted excellently for the fact that the orange and red stars are divided into giant and dwarf classes, while the hotter stars are not. The remarkable thing here is that when Lockyer put forward his ideas, the giant and dwarf divisions had not been recognized. Therefore, recognition of the classes some years later was thought to be a significant confirmation of what Lockyer had said, and the evolutionary sequence could be plotted on the H-R Diagram, as shown here.

But though the sequence seemed to be logical, it was already clear that gravitation was inadequate as an energy source. For the Sun, radiation could not continue for more than fifty million years at the very most, which was not nearly enough.

Henry Norris Russell (of H-R Diagram fame) put forward a better idea. He retained Lockyer's evolutionary sequence, making the Red Giants very young and the Red Dwarfs very old, but he introduced power from the atom, and believed that a star must shine because it was steadily converting its matter into radiation.

This is where I must say something more about the structure of the atom, though I warn you that I am going to oversimplify it inasmuch as I aim to do no more than give you a general idea of the situation. Everyone knows that matter is made up of atoms and atom-groups, but unfortunately it is impossible to give a proper description of an atom in ordinary language. Plain English will not do. To give an analogy: how can one explain music to a man who has always been completely deaf? You cannot write down a tune. You can, of course, produce a conventional representation, and here, as an example, are the first few notes of "God Save the Queen" . . .

Yet this would mean nothing at all to a person who has never heard a musical note. Our language is hopelessly inadequate.

It is the same with the atom. Many years ago scientists did give an understandable account of it, but it must not be taken literally. The best course is to describe it first and deal with its shortcomings later.

We begin with the atom of hydrogen, which is the simplest

of all (as well as being the most abundant). The hydrogen atom is made up of two parts: a central nucleus, consisting of a particle called a proton, and a much less massive particle called an electron, which revolves round the nucleus just as a planet revolves round the Sun. The proton carries a unit charge of positive electricity, while the electron carries an equal charge of negative electricity. The two electrical charges cancel each other out, so that the atom as a whole is electrically neutral. (One does not need to be an Einstein to realize that $+1 - 1 = 0$.)

Helium, the next element, is more complex. There are two planetary electrons, and since we must keep the whole atom electrically neutral we must give the nucleus a double positive charge—so that instead of being a single proton, it becomes a compound structure containing two neutrons.

We can continue up the scale, adding an extra electron for each element, and at the same time introducing an extra charge into the nucleus. Oxygen, for instance, has 16 planetary electrons. Uranium, with 92 electrons, was long thought to be the heaviest element of all, though the series has been continued recently well beyond the hundred mark. The essential point here is that there are no gaps in the sequence, and there is no room for any; one cannot have half an electron, and so it is clearly out of the question to put in a new element between, say, helium (two planetary electrons) and lithium (three).

The picture is rather like that of a miniature Solar System, with the atomic nucleus taking the place of the Sun. Moreover, the atom, like the Solar System, is mostly empty space; the nucleus and the electrons are tiny compared with the total room which the atom occupies.

To complete our picture, let us see what will happen if an atom meets with an accident and loses an electron. If we knock off one circling electron from a lithium atom, we will be left with a nucleus and two electrons—but this will not be the same as a helium atom, because the lithium nucleus can balance the electrical charge of three electrons, not two. The result will be an incomplete lithium atom with an overall positive charge. An atom which has lost an electron in this way is said to be ionized, and the process can be continued until all the planetary electrons have been stripped away, so that ionization is complete.

It all sounds very easy, but we now know that it is highly misleading to think of an atom quite in this way, and we cannot regard the protons and electrons as solid lumps. Moreover, many other fundamental particles have been discovered; in particular there are neutrons, which have no electrical charges at all. Neither do the electrons move in the same manner as planets, and it is at this point that ordinary English fails us.

Fortunately we need not go any further at the moment. Our picture of the atom as a miniature Solar System, with electrons revolving round a nucleus, is good enough for our present purpose—provided that we are careful not to take it literally. Now we can go back to Russell's idea of how a star produces its energy.

A proton, as we have seen, carries a positive charge of electricity; an electron carries an equal negative charge. Suppose that the two meet head-on, and cancel each other out? One minus one still equals nought, and Russell believed that the result would be the complete disappearance of both particles, with the production of energy. Unless the process were halted in some way or other, the whole star might eventually be turned into radiation. In any case, there would be no time-scale complications. So much energy would be available that the life-cycle of a star would extend not over a mere fifty million years, but over something more like ten million million.

Russell's annihilation of matter theory, in its various forms, held the stage for two decades after its first publication in 1913. It all seemed highly plausible. A star would begin as a Red Giant, pass down the giant branch until joining the Main Sequence at type O, B or A, and then move down the Main Sequence itself, losing mass as it annihilated its material until all its energy had been spent. As on Lockyer's original theory, the H-R Diagram showed a true evolutionary track from birth to near-death.

Then, gradually, new facts came to light, and astronomers were forced to take a long, hard look at the whole situation. It became painfully evident that whereas Lockyer's time-scale had been much too short, Russell's was much too long. Ten million million years was too extended; what was needed was a happy mean. Also, increased knowledge of atomic structure cast discredit upon the straightforward mutual annihilation of a proton with an electron. By the early 1930s the whole question was open once more, and no theory gained general acceptance. One famous astronomer said wryly that he knew all about stellar evolution in 1915, rather less in 1920, and nothing at all since 1929.

This was the situation shortly before the war, but then a sudden inspiration on the part of Dr. Hans Bethe, a German scientist working in America, opened up a new avenue. Apparently Bethe was travelling by train from Washington, where he had been attending a conference, back to Cornell University when he decided to pass the time by calculating some nuclear reactions able to account for the observed energy output of the Sun. Almost at once he hit upon a most convincing solution. We know now that he was not entirely right, inasmuch as his process is not the main one which produces the radiation of the Sun; but it is operative for many stars, and modern theories are based largely upon the Bethe formula. All in all, that particular train journey was a most important one, though it is true that rather similar work was being carried out around the same time by George Gamow in America and Carl von Weizsäcker in Germany. The key to the whole problem is hydrogen.

Hydrogen is much the most plentiful substance in the universe, and it is more abundant than all the other elements put together. The Sun, like most other stars, contains a tremendous amount of it. Near the Sun's centre, where the temperature is about 14 million degrees and the pressure is colossal, strange things are happening to the hydrogen nuclei: they band together to make helium nuclei. It takes four hydrogen nuclei to make one nucleus of helium, but Bethe saw that there must be more to it than a simple running-together, and he worked out that carbon and nitrogen, two elements very familiar to us, might act as 'catalysts'—that is to say, substances used during the series of transformations, but which themselves emerged unchanged. The net result, then, could be summarized as the steady building-up of helium out of hydrogen.

The crux of the whole theory is that the helium-building

process releases energy. The single helium nucleus produced has, moreover, slightly less mass than that of the four hydrogen nuclei which went into making it, so that mass has been lost. This is not the same thing as Russell's old idea of the direct annihilation of matter, but it gives rather the same result on a very modified scale. Each time a helium nucleus is built, the Sun loses a little mass, and energy is set free. It is this energy which keeps the Sun shining.

The carbon-nitrogen cycle is only one way in which helium may be produced from hydrogen, and it is actually another, the so-called 'proton-proton reaction', which has proved to be the more important in the case of the Sun itself. However, the final result is the same: four hydrogen nuclei produce one helium nucleus plus energy.

An atom is inconceivably small, and it is not easy to understand how such a process can keep the Sun radiating for year after year, century after century. The answer lies in the Sun's tremendous mass; there is so much hydrogen that a great deal of energy is being set free all the time. Each second, the Sun loses mass to the extent of four million tons, so that if it has taken you a quarter of an hour to read this chapter the Sun now has a mass of 3,600 million tons less than it had when you picked up the book.

I can assure you that there is no need for alarm. The loss may seem staggering, but it fits well into the time-scale. We are neither too long nor too short. The Sun has so much material that it can well afford to lose it at this rate, and there is enough 'fuel' to supply it for thousands of millions of years yet.

This is not so for the more energetic stars. Rigel, for instance, is extremely luminous, and is squandering its nuclear fuel at an amazing rate—perhaps 80,000 million tons a second. Even Rigel cannot stand this loss for long on the cosmical scale, and neither can it have existed in its present state for nearly so long as the Sun. It cannot have shone with its current brilliance for many millions of years, and it will age comparatively quickly.

Rigel, with its luminosity of around 50,000 Sun-power, is by no means the supreme searchlight of the sky. S Doradûs in the Large Magellanic Cloud, unfortunately too far south to be seen from Europe, is a million times as luminous as the Sun, and yet is so remote that with the naked eye it cannot be seen at all. S Doradûs is a real spendthrift, and it cannot have kept up this fantastic output for as much as a quarter of a million years.

It is tempting to cling to the old Lockyer sequence of evolution, making the Red Giants young, the white O and B stars middle-aged, and the dwarfs old. On this scheme, stars such as Rigel would gradually cool down into calmer bodies of the same type as the Sun, while the Sun itself would dwindle into graceful old age as a Red Dwarf. Alas, things are not so clear-cut as this, and by now we have realized that we must abandon the old Lockyer sequence entirely. It is time to take a fresh look at the whole problem, and see where more modern theories lead us.

In fact, we have already taken a preliminary glance at them; remember how our typical star begins by condensing out of nebular material, shrinks, and begins to shine by the 'convection' process. We left the story at the point where the star had definitely achieved stellar status, but no nuclear reactions had begun. So let us turn back to the H-R Diagram (Fig. 36), and start at the time when our star has begun to 'boil'. After its initial burst of glory it continues to shrink, and also becomes fainter. It is approaching the Main Sequence, and is following what is called the Hayashi track in honour of the Japanese astrophysicist Chushiro Hayashi, who first described it (in 1961, which was, incidentally, a year after the publication of the first edition of this book). From the dotted lines in the diagram, you can easily see that the original mass of the star will have a great bearing upon the Hayashi track which it follows; so for the moment let us concentrate upon the middle one, drawn for a star with a mass about equal to that of the Sun.

After a million years or so (still a very brief period on the cosmical time-scale) the star has reached the bottom of the Hayashi track, with a considerably altered internal structure. The star then evolves toward the left of the diagram along what we call the Henyey track, named after L. G. Henyey, one of the three astrophysicists who first described it. After a further ten million years, the temperature of the star's core becomes so high that nuclear reactions are triggered off—and the star falls from its Henyey track on to the Main Sequence, where it settles down to a long period of steady, stable existence.

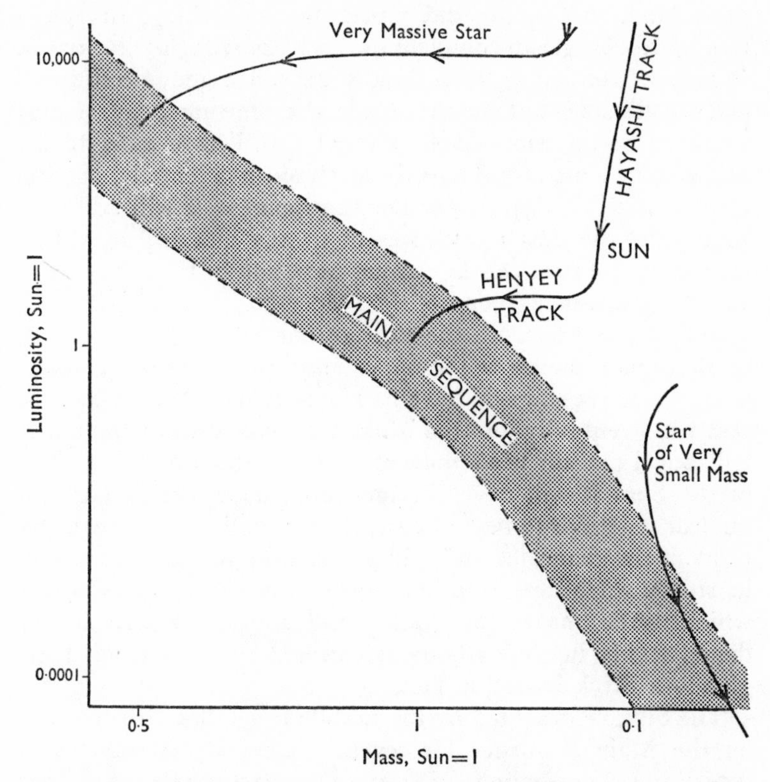

Fig. 36. Hayashi and Henyey tracks toward the Main Sequence for a very massive star (*top*), a star of solar mass (*middle*) and a star of very slight mass (*bottom*).

Young stars are generally surrounded by shrouds or cocoons of dust, so that early in their luminous careers they are trying to show themselves from behind a screen, so to speak. At some point in the story, certainly before the Main Sequence is reached, these dust-clouds are dispersed with surprising suddenness, and we have caught one or two such stars in the act. In 1936 a very faint star in or close to a dust-cloud in Orion suddenly began to brighten up, and in only 120 days it had increased from the 16th to the 10th magnitude; in other words it had become 250 times as brilliant as before, since when it has shown only slight fluctuations. In 1969 a star of the same kind, in Cygnus, did much the same thing, though it took rather longer to complete the increase (roughly 250 days). It may be, of course, that these stars had a genuine surge of extra brilliance, but on the whole the blowing-away of dust seems to be the more likely answer. I will have more to say about these stars elsewhere; their names, incidentally, are FU Orionis and V 1057 Cygni. For the moment, I will add only that while the dust is still present in any quantity it will be heated by the star, causing strong radiation in the infra-red or very long-wavelength end of the spectrum.

Before going on with the story of our solar-type star, let us pause to look briefly at the upper and lower ends of the scale. A very massive star will have its Hayashi and Henyey periods, and will eventually join the Main Sequence toward the upper left. But if the star has a mass of less than about one-tenth that of the Sun, it will never become hot enough at its core for nuclear reactions to begin at all; it will simply fall to the lower right of the diagram, and will go on shining only because of its steady shrinkage. In other words, it is an unsuccessful star which never 'makes the grade', and after a spell as a very dense, feebly luminous body it will end up as a dead, inert globe—a Black Dwarf, in fact.

The Sun, represented by the middle track, has already been on the Main Sequence for several thousands of millions of years, and even though it is using up its reserves so rapidly it will continue shining in its present state for at least six thousand million years in the future. (It is reassuring to note that the Sun has lost much less than one per cent of its total mass since the end of the last Ice Age.) But all good things must come to an end, and this applies to the Sun's stay on the Main Sequence. Eventually the supply of available hydrogen will be used up, and the central core will be made of almost pure helium. The 'fuel' has run out, and so the core contracts rapidly, though the hydrogen in the surrounding mantle is still being converted into helium as before. The shrinking of the core heats it up again, while the outer layers expand and become cooler. Our solar-type star moves to the upper right of the Main Sequence, and turns into a Red Giant. With the Sun, the diameter will then be of the order of 25,000,000 miles, and the luminosity a hundred times its present value; the central temperature will soar to something like 100 million degrees.

With yet a further increase in the core temperature will come another development. The helium itself begins to take part in reactions, and heavier and heavier elements are built up; even iron is produced. The so-called 'helium flash' is every whit as dramatic as the original start of nuclear power, but so far as the Sun is concerned we will not be able to observe it from our present vantage-point. It needs little imagination to foresee what will happen to the Earth once the Sun swells out into a Red Giant.

I do not propose to go into details of the various complicated nuclear processes which follow the helium flash; all I am trying to do is to give a general outline of the story. Finally, all the available nuclear energy has been spent. The star's period of glory is over—and it collapses, as quickly as it had expanded, moving down to the lower left of the H-R Diagram and becoming a small, incredibly dense star of the kind known as a White Dwarf. Because it is bankrupt of its reserves, and has nothing to sustain it, its atoms are crushed together, so that they are tightly packed; the density may reach over 200,000 times that of water. After another immensely long period, all light and heat leaves the erstwhile star, and it becomes a dead Black Dwarf.

As we have seen, a star with an initial mass of less than about one-tenth that of the Sun becomes a Black Dwarf without the honour of becoming a giant first, or even of joining the Main Sequence. The story of a star much more massive than the Sun is different again. It spends its stable period near the top left of the H-R Diagram, and runs through its cycle of evolution much more quickly; and it dies not with a whimper, but with a very pronounced bang—assuming that our current theories are even approximately correct. It will explode in a cataclysmic blaze which we term a supernova, and end up as a super-dense body composed of neutrons. With stars over ten times as massive as the Sun there may be a very rapid collapse, with the production of what we call a Black Hole. I will have more to say about these remarkable objects in Chapter 12.

We cannot yet pretend that we have yet worked out a really reliable picture of the whole story of a star, from formation to extinction. In particular, there is still much to be learned about the post-Main Sequence stages. Our ideas have changed out of all recognition since it was realized that nuclear reactions provide a star with its energy, and yet in one important respect our views are unaltered. Even though the stars are so long-lived, they are not eternal. Nothing in the universe lasts for ever, and there must come a time when our world is either destroyed or else survives as a scorched-up, blackened globe circling lifelessly around its dying Sun.

Chapter Nine

DOUBLE STARS

IN THE YEAR 1650 Father Riccioli, a Jesuit priest who was a keen astronomer and who became famous for drawing a reasonably accurate map of the Moon, made an interesting discovery. On turning his telescope toward Mizar or Zeta Ursæ Majoris, the second star in the tail of the Great Bear, he saw that it was double. Of course, the nearby star Alcor had been known for centuries—the pair used to be called 'Jack and his Wagon'—but Riccioli found that Mizar itself was made up of two stars, so close together that to the naked eye they appeared as one.

Riccioli did not know that the two components of Mizar are physically associated. Rather naturally, he imagined that they were simply lined up, with one member of the pair lying far beyond the other (though, in passing, Riccioli's views about the universe in general were well behind his time; he could never believe that the Earth was anything but the central body, of supreme importance). It was not until Sir William Herschel carried out his researches, more than a century after Riccioli's time, that the true nature of physically-connected or 'binary' pairs became clear. As time went by, it was seen that these binary systems are surprisingly common.

The apparent separation of the components of a double star is measured in seconds of arc. Alcor is about 700 seconds of arc from Mizar, which is rather too far for it to be classed as a 'double' in the usual sense of the term, whereas the stars of the close Mizar pair are separated by 15 seconds of arc—wide enough for the pair to be split with a small telescope, and yet not a great deal when we remember that the apparent diameter of the Moon is roughly 1,800 seconds of arc.

The 'position angle' of a double star, binary or otherwise, is the apparent direction of the fainter component (B) as reckoned from the brighter (A), beginning with 000 degrees at the north point and measuring round by east, south and west back to north (Fig. 37). Binary stars with short revolution periods have

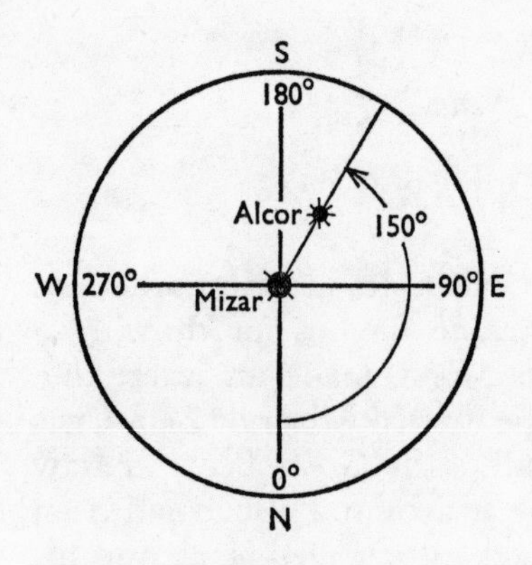

Fig. 37. Position angle of a double star. With Alcor, the position angle is 150° as reckoned from Mizar.

distances and position angles which change quite rapidly, so that it is never safe to trust the values given in books or catalogues more than a few years old. There is no such change for the optical doubles, or for physical systems whose components are at a tremendous real distance from each other.

Before saying more about binary-star astronomy, it will be useful to spend a few moments in describing some of the most conspicuous pairs. The interest of a binary system is always increased as soon as you have found it for yourself, and know what it looks like. The practical observer has a great advantage over the man who is content to study from the depths of an armchair.

Fig. 38. Theta Tauri, a naked-eye double in the Hyades, near Aldebaran

We may make a start even if no telescope is available, since in addition to the Mizar-Alcor pair there are several doubles separable with the naked eye. For instance Theta Tauri, in the Hyades star-cluster close to Aldebaran, is an easy object, and its position makes it readily identifiable (Fig. 38). In summer evenings, when the Hyades are below the horizon in Britain, we have another naked-eye double: Epsilon Lyræ, near the very brilliant Vega. And in autumn we may turn to the rather dim constellation of Capricornus, the Sea-Goat, whose main claim to fame is that it lies in the Zodiac. Both Alpha and Beta Capricorni are wide doubles (Fig. 39): Alpha or Al Giedi, with a separation of 376 seconds of arc, needs no optical aid at all, and is convenient enough to lie in line with the trio of stars in Aquila of which Altair is the leading member. Beta Capricorni has components of magnitudes $3\frac{1}{4}$ and 6, separated by 205 seconds of arc, so that binoculars will deal adequately with it.

A small telescope gives enormous scope. Quite the loveliest

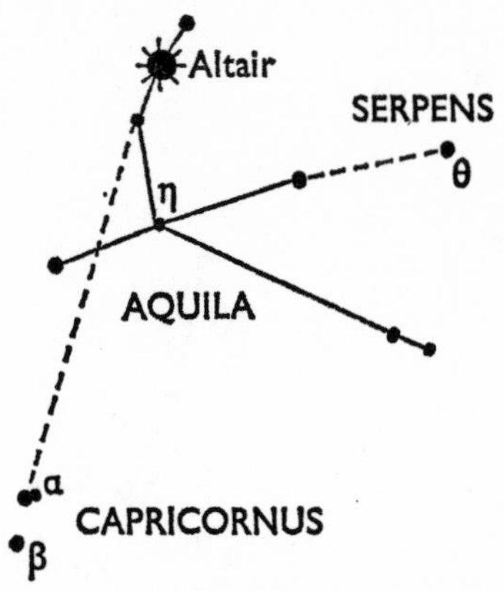

Fig. 39. Position of the naked-eye double star Alpha Capricorni.

double of all is Albireo or Beta Cygni, the faintest star of the cross of the Swan. It is of the third magnitude, and is easy to find (Fig. 40); for one thing it lies very roughly between Vega and Altair. The golden-yellow primary is attended by a fifth-magnitude companion which some people call green and others blue. The colours are vivid, and are enhanced by contrast. Even good binoculars will split the pair. Though they form a true system, the components are so far from each other that they seem to all intents and purposes fixed in separation and position angle, though they share a common motion through space.

Very different is Theta Serpentis or Alya, which lies not far from Altair, in line with the three stars of the Eagle which are lettered Delta, Eta and Theta Aquilæ. (Eta Aquilæ, in passing, is a famous variable, of which more anon.) In Theta Serpentis we have two stars, each of magnitude 4½, which seem to be perfect twins, alike in every way, so that it is impossible to say which is A and which is B. With Gamma Virginis or Arich, in the 'Y' of Virgo between Spica and Regulus (Fig. 41), the two components are also almost equal. The revolution period is 180 years, so that the aspect changes markedly over the course of a lifetime. The separation was at its widest (just over 6 seconds of arc) in 1920, but is now closing, and by the minimum separation, in the year 2016, Arich will appear single except in very large telescopes. When I started looking at double stars with the aid of a small telescope, around 1930, I always regarded Arich as one of the

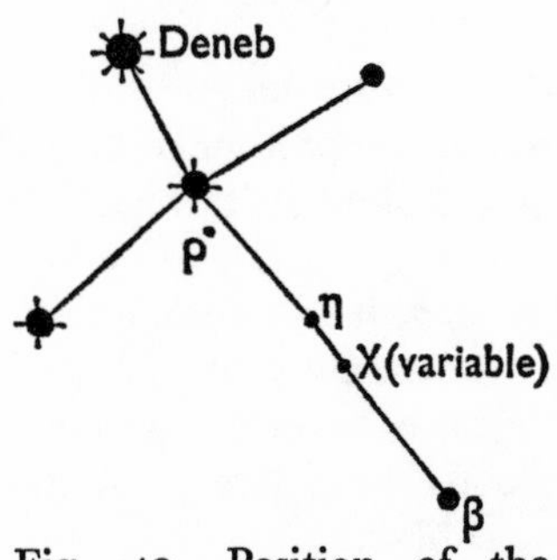

Fig. 40. Position of the double star Beta Cygni.

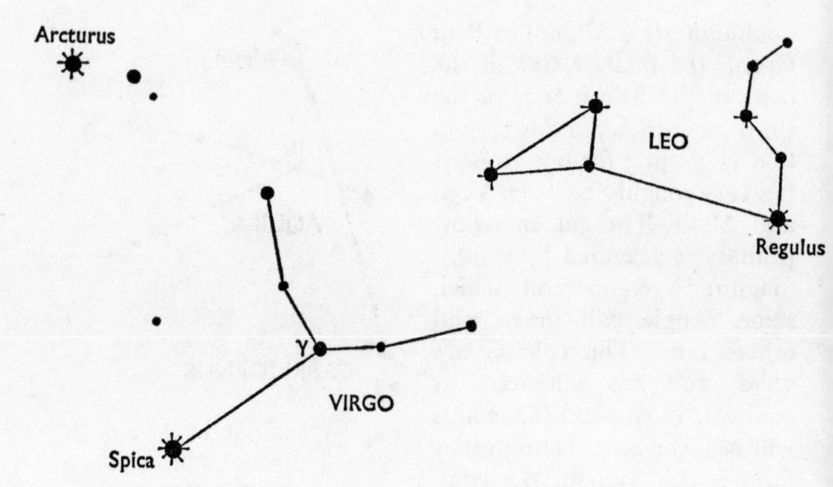

Fig. 41. Position of Gamma Virginis (Arich). It is easily found, at the junction of the Y of Virgo.

'show-pieces' of the sky, but it is not nearly so spectacular now.

Another case of virtual twinning is Gamma Arietis (Fig. 42), in the Ram, not far from the Square of Pegasus. And there is a difference of about a magnitude between the two main components of Castor in Gemini; here again the separation is much less than it used to be. The period of revolution is 350 years, and Castor is not now a very easy pair with small apertures.

Southern observers have two magnificent binaries in Alpha Centauri, the closest of the bright stars, and Acrux, leader of the Southern Cross. There are many other examples too, but I will not go into further details now, as I have done so elsewhere.*

With some doubles, one component is very much brighter than the other. The brilliant Rigel has a companion of below the sixth magnitude; a small telescope will show it, though it is very much overpowered by the dazzling radiance of the primary. The Pole Star has a ninth-magnitude companion, visible with a 3-inch refractor. Antares in Scorpio and Rasalgethi in Hercules, both vast M-type supergiants, have greenish companions; with Epsilon Boötis, near Arcturus, the primary is yellowish and the companion blue.

These are just a few of the many double stars in the sky. The

* In my book *The Amateur Astronomer* (Lutterworth Press, 1974).

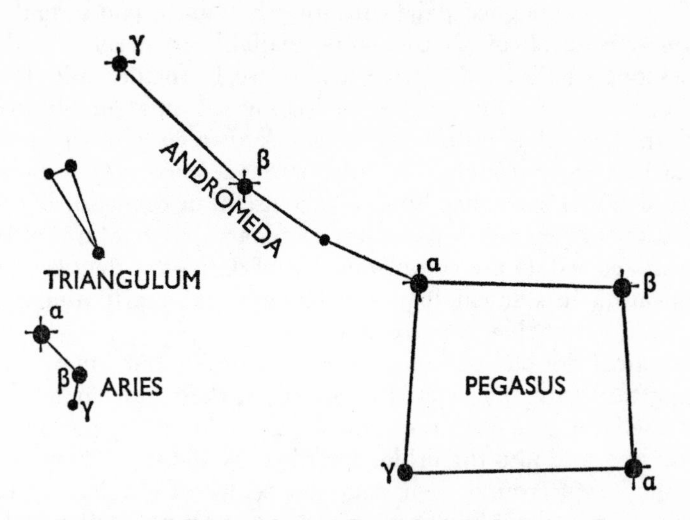

Fig. 42. Position of the double star Gamma Arietis. It is only of the fourth magnitude, but it is easy to locate.

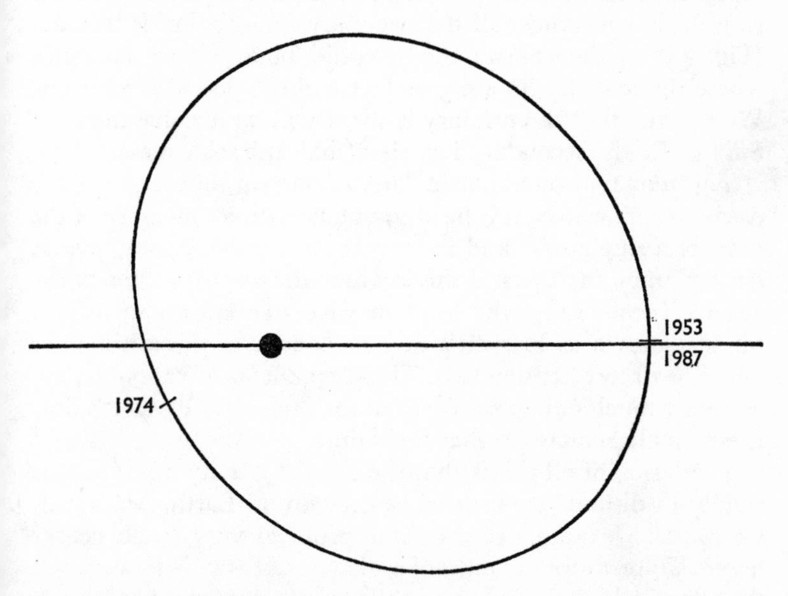

Fig. 43. Simplified apparent orbit of the fine binary Zeta Herculis, where the period is 34 years.

full list of catalogued pairs runs into thousands, and to make a complete check of all the pairs available to even a modest telescope would need a great deal of work. Incidentally, there is scope here for the serious, well-equipped amateur observer. By the use of a measuring device known as a micrometer, attached to a telescope of adequate aperture (say a 6-inch refractor) he can make himself very useful in re-measuring the position angles and separations of binary pairs. Many of the published values are completely out of date—yet they keep on appearing in star catalogues, textbooks and yearly almanacs. It is time that they were revised.

Optical doubles are of no special interest, but binaries are important in many ways. In particular, their movements give us a reliable clue as to their masses. If we know the parallax of a binary, and also the orbits described by the two components as they move round their common centre of gravity, we can work out the combined mass of the system compared with that of the Sun; and if we can measure the orbit of either component separately, we can find the individual masses of the two stars as well. For instance, all the necessary information is to hand (Fig. 43) in the case of the splendid binary Zeta Herculis, where the magnitudes are 3 and 6½ and the period is 34 years. We can tell that the primary is slightly more massive than the Sun, while the secondary has about half the solar mass.

This may not sound particularly important, but actually it is of vital significance. It is hard to obtain a direct measure of the mass of a single star, and in fact we cannot do so; we have to depend upon theory, and this is where the binaries come to the rescue. Incidentally, the most massive star known to us is a binary, known as Plaskett's Star in honour of the astronomer who first drew attention to it. The components are about equal, and very much more massive than the Sun—but Plaskett's Star is extremely remote, so that it is faint.

The lesson of all this is that the laws of gravity apply just as rigidly in distant star-systems as they do on Earth. Moreover, we have an example of the same problem very much nearer home. To measure the mass of a planet, one way is to work out the effects which it produces in the movements of other bodies. Up to 1877 Mars, which is a small world, was believed to be moonless, and its mass was not easy to find, since its pull upon

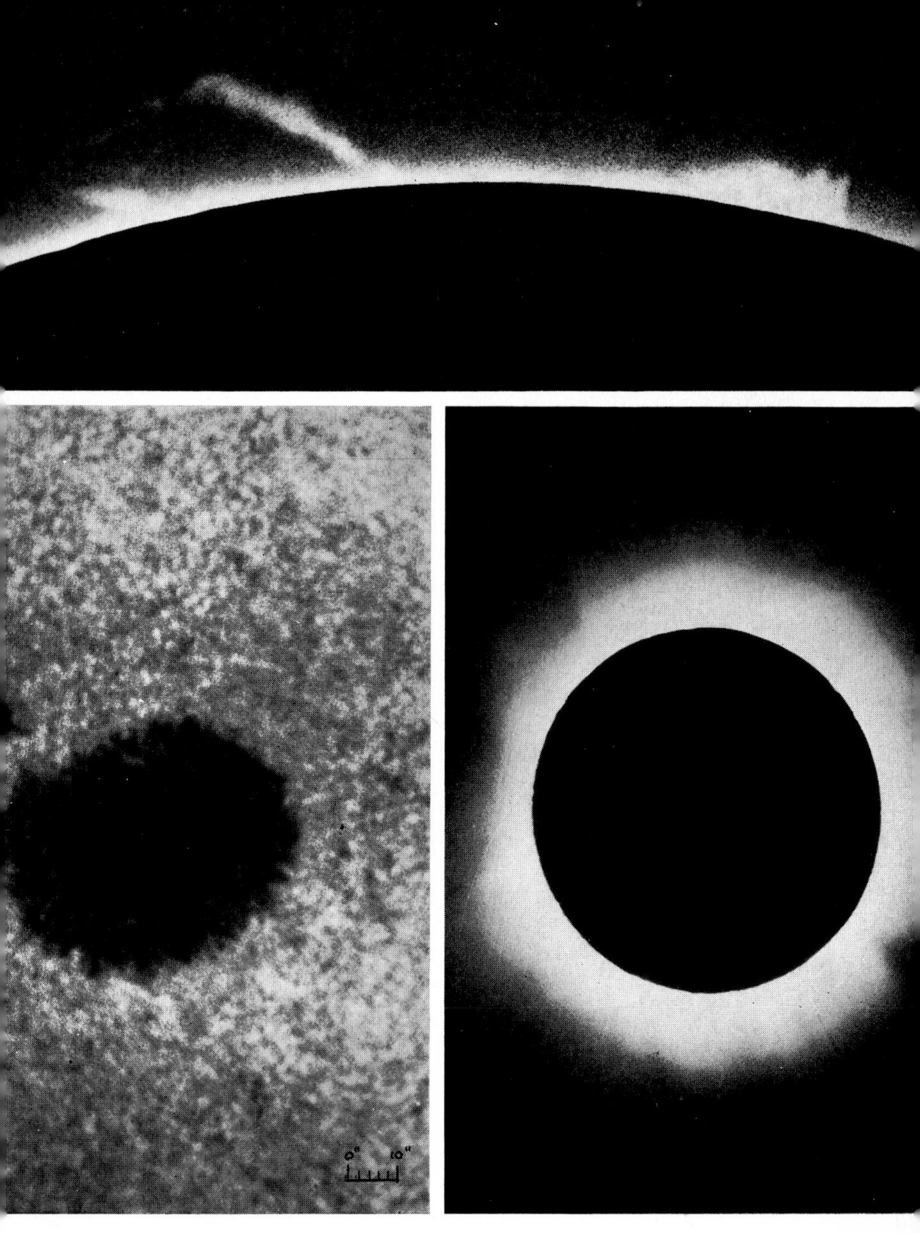

I. *The Sun*

(*top*) Prominences, photographed on 27 March 1971, at 11h 13m, by Ramon Lane.
(*lower left*) Huge sun-spot, photographed on 27 Sept. 1970, 08.00h, by W. M. Baxter: 4in. O.G.
(*lower right*) Total eclipse of the Sun: 30 June 1973, Dr. R. Maddison, North Africa.

II. *Nova Persei*

Expanding nebulosity round Nova Persei (GK Persei) 1901. 200in. reflector, photograph from the Mount Wilson and Palomar observatories. This picture was taken half a century after the 1901 outburst!

III. *Supernova*

Supernova in NGC 7331. Lick Observatory photographs. The supernova is not shown on the upper picture; it is arrowed in the lower photograph, which was taken when the supernova was at its brightest.

IV. *Double cluster*

The 'Sword-Handle': the double cluster in Perseus; χ-h (H.VI.33–4); it is not included in Messier's catalogue. Lick Observatory photograph.

V. *Globular Cluster*

Messier 5 (NGC 5904), the globular cluster in Serpens. Distance 27,000 light-years. This cluster is large, bright and easily resolved into stars.

VI. *Nebulosity*

Bright and dark nebulosity near the star Gamma Cygni. 48in. Schmidt, Palomar. Photograph from the Mount Wilson and Palomar Observatories.

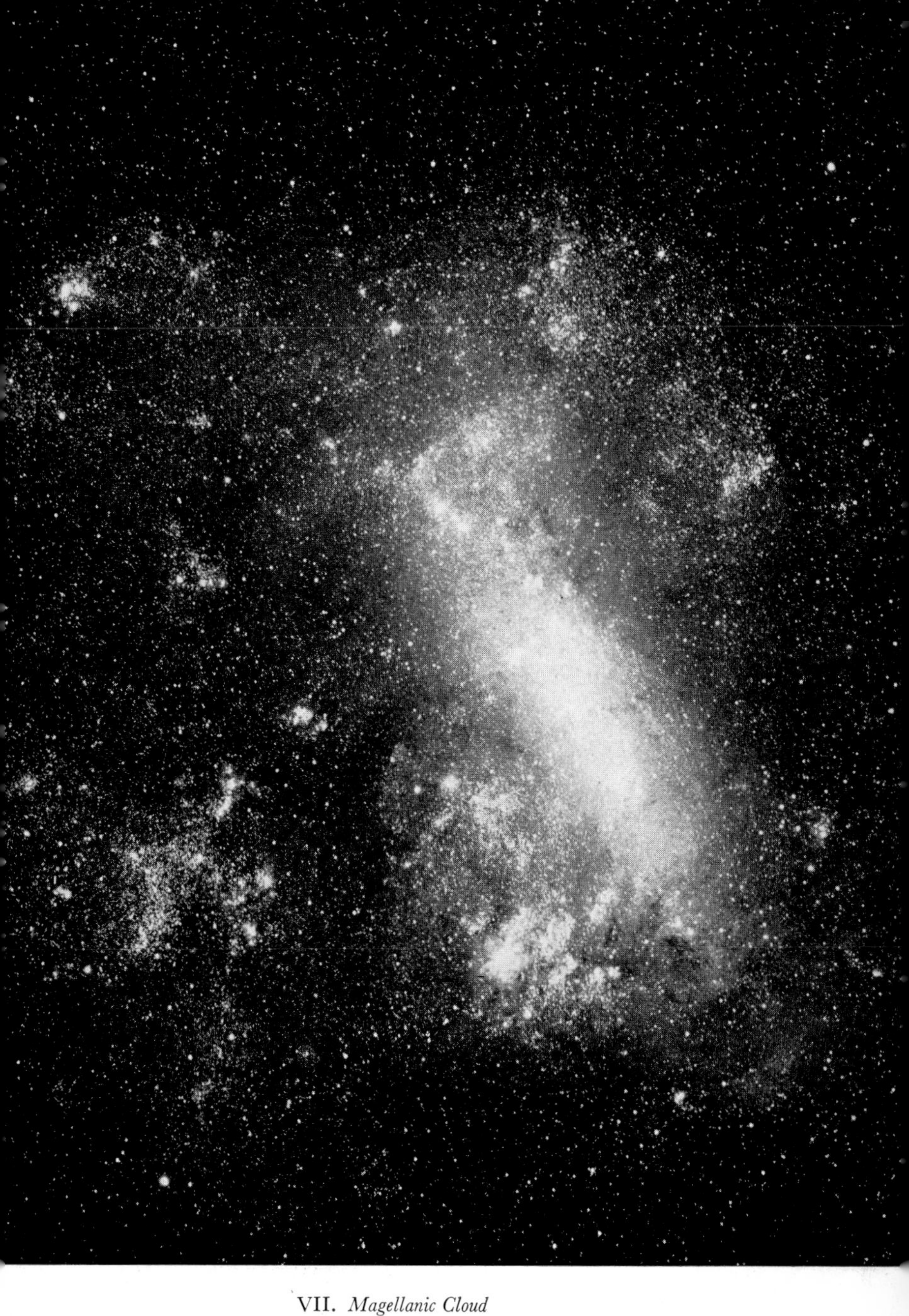

VII. *Magellanic Cloud*

The Large Magellanic Cloud (Nubecula Major)—the brightest of the external systems, but unfortunately too far south to be seen from Europe.

VIII. *Triangulum Spiral*

M.33 (NGC 598), the spiral galaxy in Triangulum, photographed in red light with the Palomar 48in. Schmidt. Photograph from the Mount Wilson and Palomar Observatories. M.33 is a member of the Local Group.

IX. *Cluster of galaxies in Hercules*

Cluster of galaxies in Hercules. 200-in. reflector, Palomar. Photograph from the Mount Wilson and Palomar Observatories. Galaxies of all types are shown, together with foreground stars of our own Galaxy.

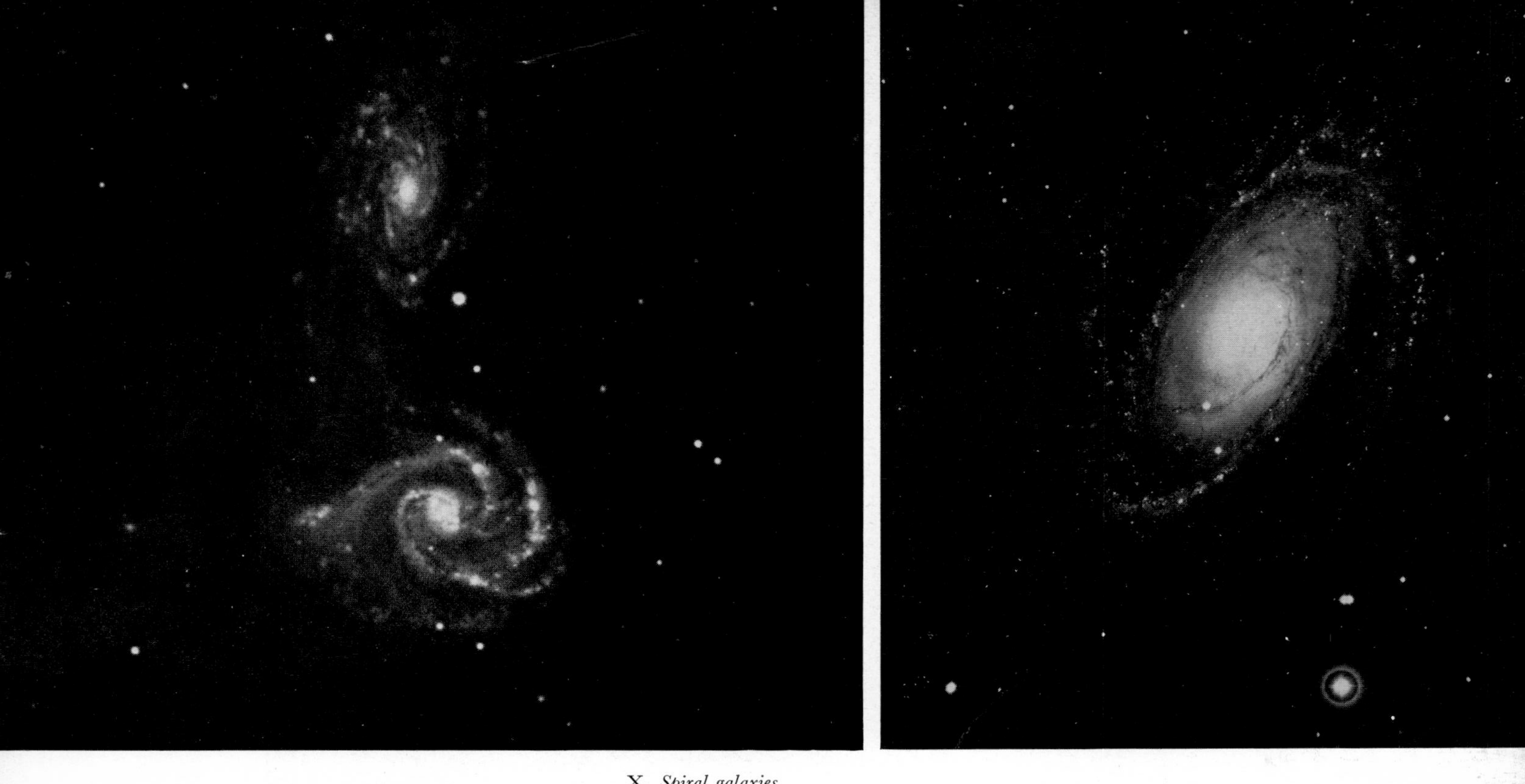

X. *Spiral galaxies*

(*left*) NGC 5432 and 5435; intertwined spirals (Lick Observatory photograph).
(*right*) Messier 81 (NGC 3031), a spiral galaxy in Ursa Major; it is 8·5 million light-years away, and is the chief member of a bright group of galaxies.

XI. *M.87 Galaxy*

Messier 87 (NGC 4486). The giant elliptical galaxy in Virgo: distance 41 million light-years. The curious 'jet' is not shown here. M.87 is a strong radio source.

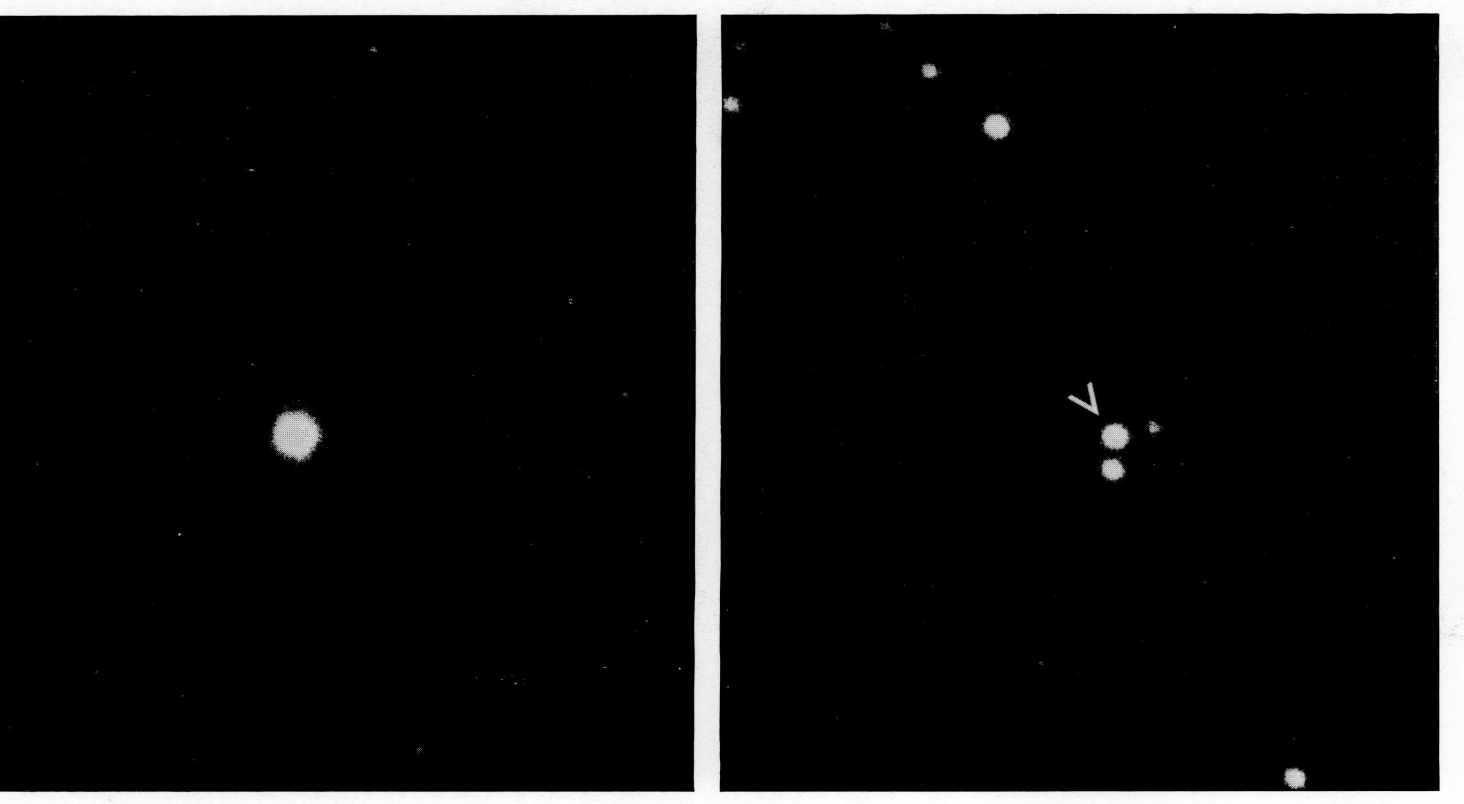

XII. *Quasars*

(*left*) 3C–48;
(*right*) 3C–147. Though quasars look superficially very like stars, their true nature is very different.

the Earth and other planets is relatively feeble. Then Asaph Hall, in Washington, found two tiny satellites, Phobos and Deimos, which whirl round Mars at distances of only a few thousand miles. As soon as the orbits of these moonlets had been worked out, which took only a few days, the mass of Mars itself could be determined. In the same way, we can find the mass of a star much more easily if there is a companion star nearby upon which the gravitational effects may be measured.

When we look at the spectral types of binary stars, we find some interesting laws. For almost identical twins, such as Gamma Virginis, the spectra are nearly always similar. When there is a marked difference in brightness between the two components, the spectra too are different. If both stars belong to the Main Sequence, the companion is generally of later type than the primary; if the primary is a giant, the companion is either a giant of earlier spectral type or else a dwarf of similar class. (Though the sequence O, B, A . . . M is no longer thought to be evolutionary, we still term the hot stars 'early' and the red ones 'late' in spectral types; for instance Arcturus, of type K, is said to be of later type than Sirius, whose spectrum is of type A.) Let me stress, however, that the law is not rigid, and not all binaries obey it.

How were binaries formed? It is tempting to suppose that a system of this kind is due to the break-up of an originally single star, because of excessively rapid rotation, but nowadays this attractive idea has fallen into disfavour, and it is usually thought that the two members of a binary were formed close together, so that they have remained linked.

So far I have been talking about binaries in which both components are visible in small telescopes. This is not, however, the case with one of the most remarkable binaries in the sky: Sirius, the Dog-Star.

Sirius shines so brilliantly that it is hard to realize that it is by no means a supergiant. True, it is 26 times more luminous than the Sun, but it owes its great brightness to the fact that it is less than nine light-years away; of the first-magnitude stars, only Alpha Centauri is closer. Naturally, Sirius has an unusually large proper motion, which is why Edmond Halley was able to show that it had moved perceptibly across the sky since the days of Hipparchus and Ptolemy.

In the ordinary way, a star with large proper motion travels across the sky in a uniform manner, but in 1834 Friedrich Bessel—later to achieve extra fame as being the first man to measure the distance of a star—realized that Sirius was behaving oddly. Its motion was erratic, so that instead of travelling in a direct line it was 'waving' its way along. Each 'wave' was very tiny, and took about fifty years to complete, but the effect was certainly there.

No single star could possibly act in such a fashion, and Bessel therefore suggested that Sirius must have a binary companion, too faint to be seen and yet massive enough to pull on the bright star and account for the waving motion (Fig. 44). Bessel reached this conclusion not long before his death, in 1844, which is interesting because research of rather the same kind was

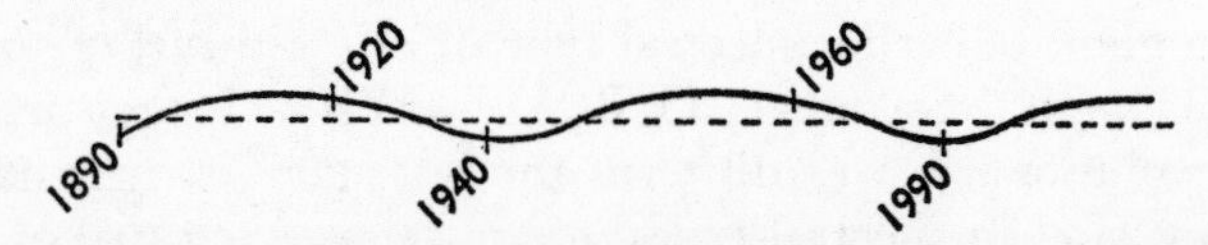

Fig. 44. The 'wobbling' of Sirius as it moves against the background of more distant stars; the positions are shown to either side of the average track. The perturbations are due to the White Dwarf companion.

going on in connection with our own Solar System. The planet Uranus, discovered by Herschel in 1781, had been causing trouble, as it obstinately refused to move according to the path which had been worked out for it. Two mathematicians, Adams in England and Le Verrier in France, had decided that it must be affected by an unknown planet, and by studying the wanderings of Uranus they managed to track down the disturbing body—the planet which we now call Neptune. The crux of the matter was that Neptune's position was worked out before astronomers actually started to look for it with their telescopes. In the same way, the position of the binary companion to Sirius was calculated, but the faint star refused to show itself.

In 1862 Clark, a well-known American telescope-maker, was testing a new refractor with a 20-inch object-glass. He turned it toward Sirius, and saw a dim dot of light beside the brilliant star. The dot was, of course, the long-awaited Companion, just where Bessel had predicted. It is known officially as Sirius B, and unofficially as the Pup, since Sirius itself is the Dog-Star.

The magnitude of the Pup is 8½, which means that if it could be seen on its own it would be easy enough in a small telescope, but it is so overpowered by the light of the primary that it is not at all obvious. The position angle and separation alter fairly quickly, as the period is only 50 years; the distance will be at its greatest (11·5 seconds of arc) in 1975. It has been claimed that the Companion may be glimpsed with a 6-inch telescope, but I have never yet seen it with my own 12½-inch reflector, mainly because Sirius never rises high above the British horizon.

When Clark made his discovery, the 'Sirius riddle' was regarded as closed. The Companion had been expected, and it had been duly found, which was most satisfactory. It was ten magnitudes or about 10,000 times fainter than the primary, and had presumably about 1/360 the luminosity of the Sun. It was assumed to be a cool, red star of late spectral type.

Then, in 1915, W. S. Adams, of the Mount Wilson Observatory in California, produced an astronomical bombshell. He was able to study the spectrum of the Companion, and to say that the results were unexpected would be to put it mildly. Instead of being cool and red, the Pup turned out to have a curious spectrum indicating a white star with a surface temperature of 8,000 degrees, as against the modest 6,000 degrees of the Sun. The supposedly dim red object had revealed itself as extremely hot and very white.

Astronomers were faced with a set of facts which seemed to add up to pure nonsense. The mass of the Pup, as shown from the combined movements of it and the bright star, was very nearly equal to the Sun's. The temperature measurement showed that each square inch of the surface was radiating roughly 3¾ times more light and heat energy than a square inch of the solar surface. Therefore, the surface area of the Sun had to be 360 × 3¾ times that of the Pup—which led to the conclusion that the diameter of the Pup must be only 24,000 miles, a mere three times as great as that of the Earth and much less than that of a large planet such as Jupiter, Saturn or Uranus (Fig. 45).

Now let us check the density of the Pup material. What we

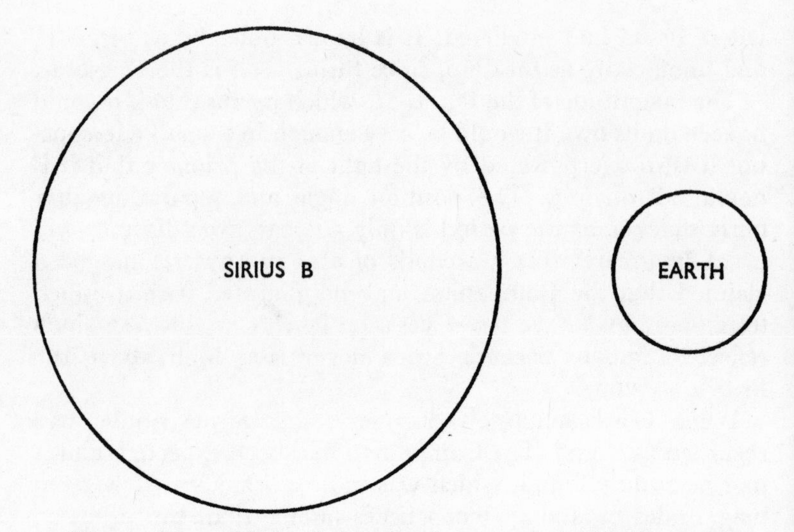

Fig. 45. Size of Sirius B (the Pup), compared with the Earth.

have to do is to pack the mass of the Sun into a globe 24,000 miles across (remembering that the Sun itself has a diameter of 865,000 miles). The only way to do so is to make the material very dense, and the calculated value turns out to be 70,000 times the density of water, so that if you could take a matchbox and fill it with matter of this kind the total weight would be more than a ton. Near the centre of the star the density would be greater still, and you could pack perhaps 50 tons into your matchbox.

It is hardly surprising that at first astronomers refused to accept any such value. It seemed quite incredible, and the easy way out was to suppose that some mistake had been made in the measurement of the temperature. Yet the facts could not be denied. The Pup was certainly faint, and it was certainly hot, so that the only solution was to give it an extremely small diameter. Gradually, the puzzled scientists came to the conclusion that 'super-dense' material must exist after all.

Were there any other stars of the same kind? This was a question of vital importance, and before long it was answered. Instead of being a freak, the Pup represented a very common class of star. Nowadays we call them White Dwarfs; they lie to the lower left of the H-R Diagram, and there can be no doubt that they are stars which have used up their nuclear energy, so that they are bankrupt of their reserves. The Pup is by no means the extreme example. Kuiper's Star (or, to give it its official designation, A.C.+70°8247) is no larger than Mars, with a diameter of about 4,000 miles, but its mass is the same as the Sun's. The material is incredibly heavy, so that if we could take a cube of it with each side of the cube measuring one-tenth of an inch, and bring it to the Earth, the weight would be about half a ton.

White Dwarfs were mentioned in the last chapter, and I stressed then that their densities are so great because the atoms in them are broken up or ionized—and hence can be packed together with very little waste space. Material of this sort is termed 'degenerate'. One way to explain what is meant is to go back to the admittedly misleading analogy between an atom and the Solar System. If you took the Sun and the planets individually and forced them close together, they would take up much less room than the Solar System actually does, because the orbit of the outermost planet has a radius of over 3,000 million miles; the proportion of wasted room in the Solar System is about the same as for an ordinary atom, but in a White Dwarf the components are crammed together. (Remember, though, that it is dangerous to take this sort of comparison at all literally, since we cannot regard the particles in an atom as solid lumps.)

Before leaving the Sirius pair, I cannot resist saying something about a problem which has always fascinated me. The spectrum of the primary is of Type A, and the colour is pure white; the rainbow flashes seen when the star is low down are due solely to the effects of the Earth's atmosphere. Yet Ptolemy, in the second century A.D., distinctly stated that he saw Sirius as red; so did some of the ancient Egyptians and Assyrians. Ptolemy was a careful and reliable observer; from his home in Alexandria, Sirius rises much higher than it ever does in Britain, so that the twinkling is less violent. Certainly Sirius is not red today, and it has been colourless ever since the tenth century A.D., when it was described by the great Arab astronomer Al-Sûfi.

This is a true puzzle. There seems no reason why Ptolemy and the other ancient observers could make so obvious a mistake. Sirius itself cannot have been red two thousand years ago; there is no conceivable way in which a red star could change into a Main Sequence star of Type A. The fascinating suggestion has therefore been made that it is the Pup which is responsible, and that it used to be a Red Giant.

Outwardly this may sound plausible, and it is true that if our modern ideas are correct the Pup must have gone through the Red Giant stage before collapsing into a White Dwarf. Unfortunately there are major difficulties. The time-scale is wrong, and a Red Giant shining together with the present-day Sirius would make an object as bright as Venus, which is equally at odds with the old descriptions. My own view, for what it is worth, is that we must admit an error by Ptolemy. All the same, it is a curious episode which adds even greater interest to the Dog-Star and its Pup.

I have wandered some way from the main theme of this chapter. Since the White Dwarfs, as a class, first became known because of the Companion of Sirius, a digression has been permissible, but it is time to return to our pairs of stars.

If the two components of a binary are extremely close together, no telescope will split them. Here, as is so usual in stellar astronomy, we can make use of the spectroscope, which makes it possible for us to study binary systems which appear telescopically as single specks of light.

Suppose that we are dealing with two stars (Fig. 46), very near each other and of much the same luminosity. In this case the period of revolution will presumably be short: less than a month, since if the period were greater the separation would also be greater, and we would have an ordinary visual pair. If the orbital plane is more or less edge-on to us, there will be times when one star (A) is approaching us while the second star (B) is receding; remember that the centre of gravity of the system will be roughly midway between the two. This is shown in the first diagram. In the second diagram, the two components are almost in the same line of sight, and the motions will be transverse relative to the Earth. (Of course, all these movements are superimposed on the general towards-or-away motion of the whole binary system with respect to the Earth, but this can easily be taken into account.)

In Position 1, A will have a violet shift in its spectrum, while

B will have a Doppler shift to the red. In Position 2, the spectral lines of A will be superimposed on those of B, because there will be no Doppler shifts at all, and the result will be a spectrum which appears perfectly normal in so far as the line positions are concerned. If, therefore, we come across a star

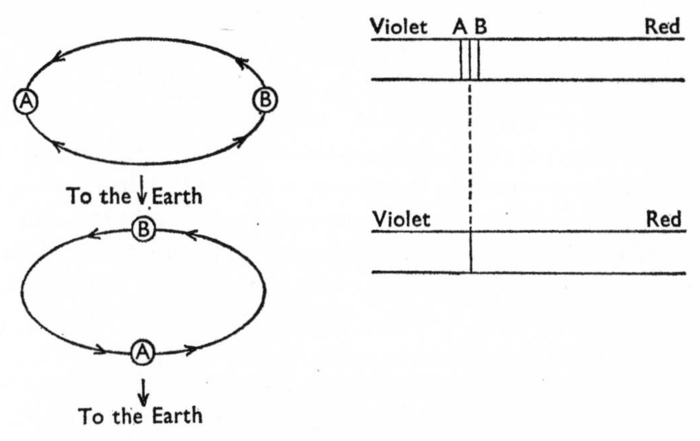

Fig. 46. A spectroscopic binary. (*Upper*) Star A is approaching the Earth, and shows a violet shift in its spectrum; B is receding, and shows a red shift, so that the combined spectrum shows a double line. (*Lower*) The motion is transverse, and the line appears single.

whose spectral lines are periodically doubled, we may be sure that we are dealing with a very close binary. Such is the brighter component of the Mizar pair, as E. C. Pickering found as long ago as 1889. The lines appear double at regular intervals, and it is found that the orbital period of the 'twins' is 20½ days. Another example is Beta Aurigæ, near the brilliant Capella, also studied in 1889. Thousands of these spectroscopic binaries are now known.

Capella is a case in point. In ordinary telescopes it appears single, but the spectroscope shows it to be a binary; the components are about 80 million miles apart, and have an orbital period of just over 100 days. One star is of type G, with a mass 4½ times that of the Sun; its F5-type companion has 3½ times the Sun's mass. Modern instruments have revealed Capella as a visual binary, though of course there are few instruments powerful enough to split it. Up to now the spectroscopic binary with the shortest known period is AM Camelopardalis, in the obscure constellation of the Giraffe. The orbital time is a mere 17½ minutes!

We also know of many cases in which the spectrum of the companion star is too faint to be seen at all. This means that the lines due to the primary will oscillate to and fro around their mean position—to the violet side when the primary is approaching, to the red side when it is receding.

Binary stars have proved to be so common that some astronomers regard them as the rule rather than the exception. Less common, though still reasonably frequent, are 'family parties' of stars—triple, quadruple and so on. Mizar is a good example. As we have noted, it is a visual binary, and the brighter component is itself a spectroscopic binary, so that we have three stars in association. But an even finer example of a multiple star is Epsilon Lyræ, not far from Vega.

Epsilon Lyræ is a naked-eye double, and people with normal eyesight will have no trouble in splitting it when the sky is dark and clear. The distance between the two components is 208 seconds of arc. A 3-inch telescope will show that each component is again double, so that we have a true quadruple system (Fig. 47). All four stars are similar in luminosity and spectral type; all are considerably more powerful than the Sun.

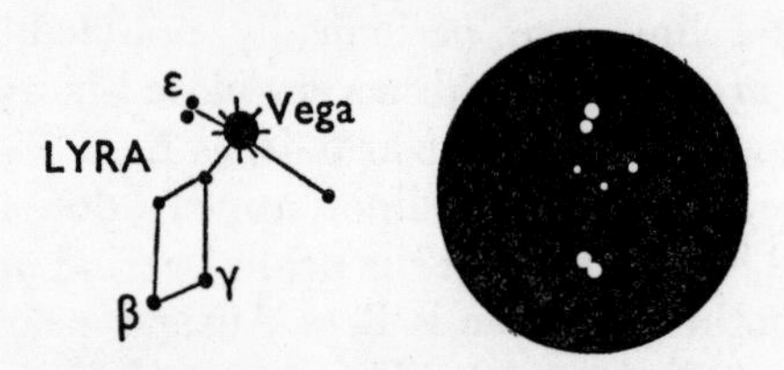

Fig. 47. Position of the multiple star Epsilon Lyræ, close to Vega. (*Inset*) The group as seen with a moderate power on a 3-inch refractor.

The movements of the stars in the Epsilon Lyræ system are decidedly complex. Each close pair has a revolution period of several hundreds of years, but the two pairs take an immense time to complete one circuit round their common centre of gravity. In fact, no measurable shift has been found, so that the period may well amount to more than a million years, and all we can really say is that there is a definite physical connection. It is always worth looking at Epsilon Lyræ, since any modest instrument will show it well.

Another famous multiple is Castor, the senior but fainter member of the two Twins. The main star is a binary; the two components differ in brightness by about a magnitude, and the period is 350 years. Each star is a spectroscopic binary, the periods being three and nine days respectively. At a distance of 73 seconds of arc lies a third member of the system—Castor C (sometimes called YY Geminorum), made up of two Red Dwarfs moving round each other in 19 hours. The Red Dwarf pair certainly orbits the main quadruplets, but must take millions of years to do so.

This is a real stellar family. Castor is not the single dot of light which appears to the naked eye; it is made up of six separate suns, four brilliant and two dim, arranged in pairs. It is logical to assume that all six are of about the same age, and were formed by the same process, but they have had very different life-stories.

Yet another interesting system—a triple, this time—is that of Alpha Centauri. The main star is a superb visual binary, and Proxima, which has the distinction of being the nearest star to the Solar System, is also a member of the family, as is shown by the fact that it has the same motion through space. Proxima is a celestial glow-worm; despite its nearness (only 4·2 light-years) it shines as a star of only magnitude 10·7.

For other examples of multiple stars we can turn back to Orion, that magnificent constellation which seems to be able to supply us with anything we may want in the nature of stellar wonders. In the gaseous nebula which marks the Hunter's Sword we find Theta Orionis, known as the Trapezium because of the arrangement of its four chief components. Sigma Orionis, between the Nebula and the stars of the Belt, is another multiple, less striking than the Trapezium but well worth examination.

There is a great fascination in these pairs and groups of stars. Wide or close, optical or binary, coloured or colourless, they can provide the casual observer with endless hours of enjoyment, and they also provide the well-equipped amateur with the chance to do some extremely useful research.

What if we lived on a planet circling round a binary star?

We would have two suns instead of one. There might be a huge Red Giant accompanied by a dim but massive White Dwarf, or alternatively we might be treated to the spectacle of a yellow sun with a blue companion, which would certainly provide us with colour effects beyond the wildest dreams of even a surrealist artist. Whether such binary systems do have families of planets remains to be seen. Meantime we must be content with our own single sun, which sends us so much radiation that we do not always appreciate that it is nothing more than one of the Galaxy's Yellow Dwarfs.

Chapter Ten

VARIABLE STARS

ON THE EVENING of 12 November, 1782, an eighteen-year-old boy named John Goodricke was busy observing the stars. He was a serious astronomer, but an unusual one; he was deaf and dumb, and had been from birth, though there was nothing the matter with either his eyesight or his brain.

Goodricke was particularly interested in Algol, or Beta Persei. In the ordinary way Algol is about as bright as the Pole Star, but on this particular occasion something was happening to it; it was fading, and continued to do so for several hours until it had dropped to magnitude 3½. Then, after less than half an hour, it began to increase once more until it had regained its former brightness.

Goodricke was not the first to observe this curious behaviour in Algol. Geminano Montanari, an Italian professor of mathematics, had noticed it as long ago as 1669. Yet nobody had explained it satisfactorily, and Algol was obviously different in type from the other known 'variable star', Mira in the constellation of Cetus. Goodricke believed that he had the answer, and in the following year he wrote a paper in which he suggested that the changes in brightness shown by Algol were due to the periodical eclipse of the bright star by a fainter binary companion.

We now know that Goodricke was right, and it was a tragedy that this gifted deaf-mute should have died soon afterwards at the early age of twenty-one. Had he lived, he would have done much for astronomical science.

Algol is easy to find. The best pointer to the constellation Perseus is Cassiopeia; an imaginary line from two of the stars in the W leads on to Mirphak or Alpha Persei, of the second magnitude, and Algol is not far off (Fig. 48). Moreover, it is very prominent when at maximum, as it is for most of the time. For 2 days 11 hours it shines steadily (or virtually so), but then fades down to minimum in only five hours. After minimum, a further five hours is needed for return to maximum,

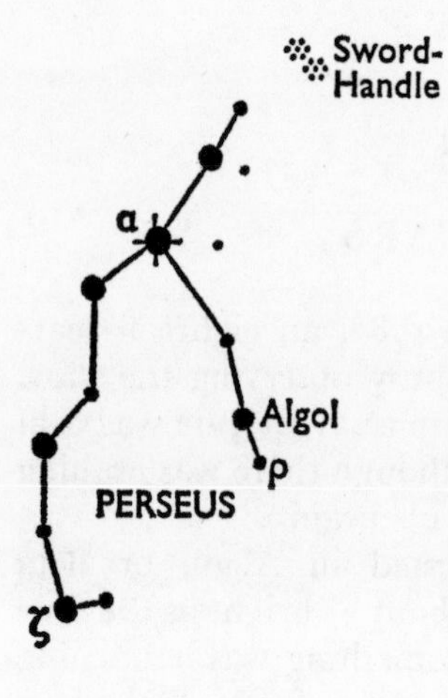

Fig. 48. Position of Algol, which normally shines as a star of the second magnitude.

after which nothing much happens for a further 2 days 11 hours. In fact, Algol 'winks' regularly.

The best way to show the behaviour of Algol, or any other star which changes in brightness, is by means of a light-curve, in which magnitude is plotted against time. It is worth noting the very slight secondary minimum between the two main winks; this is caused by the eclipse of the fainter component by the brighter, and is too slight to be detected without sensitive equipment.

There is a rather interesting point about the name 'Algol', which is Arabic, and means 'the head of a ghul'—a ghul being a female demon with an unprepossessing appearance and even more unprepossessing habits. In classical legend the star is equally sinister, since it marks the severed head of the Gorgon, Medusa, still carried by Perseus in his journey across the sky. This has led to the suggestion that the Arabs of a thousand years ago knew about Algol's strange behaviour. Modern scholars tend to disagree, and it is not now thought that the variability was known before Montanari discovered it in 1669, but it is certainly appropriate that Medusa's head should be marked by a winking star.

Goodricke's suggestion sounded reasonable enough when he first made it, but for a long time there was no proof. Sir William Herschel made a special point of examining Algol carefully with his largest telescopes, but could never see it as anything but a single point. This is not surprising, since we now know that the distance between the two components of the binary is not much more than six million miles; the angular separation is too small for the components to be seen individually, and so, as usual, we turn back to the spectroscope.

It appears that the Algol system is very complex. The brightest member is of spectral type B8, with a surface temperature of 12,000 degrees and a diameter of about 2½ million miles. The secondary is larger but dimmer; its spectrum is of later type, and the surface temperature is appreciably less than the Sun's, while the diameter is three million miles. Obviously, then, the principal 'wink' observed by Goodricke takes place when the secondary (B) passes in front of A; the smaller minimum when B is hidden by A was too slight for him to detect. Actually, the eclipses are not total, as the next diagram shows (Fig. 49).

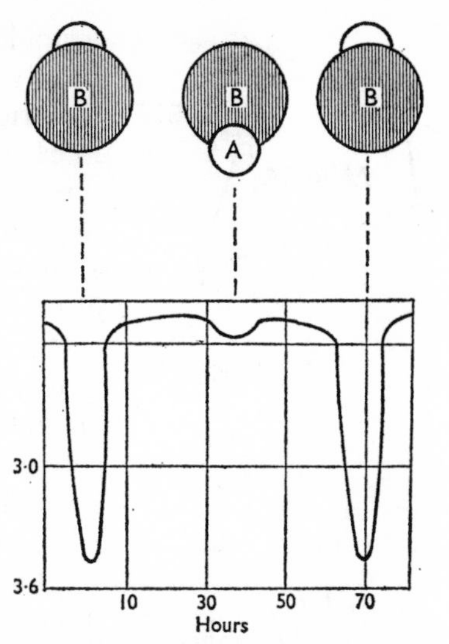

Fig. 49. Light-curve; the small 'wink' is too slight to be noticed with the naked eye. Reason for the 'winks'; when the smaller, brighter star is eclipsed by the larger, fainter component, there is a principal minimum. With Algol, the eclipse is not total as seen from Earth.

Algol A and B move together with respect to a more distant component, Algol C; the orbital period is about 1 year 9 months. C has never been seen visually, but its spectrum can be observed, and proves to be of either late A or early F type, so that the star is more luminous than Algol B. It is thought that there may be a fourth member of the system which moves round the bright binary in a period of about 188½ years.

I have already discussed spectroscopic binaries—Mizar is a good example—and in point of fact the only difference is that with Algol, the orbit is tilted so that we see its plane almost edge-on. It is not truly variable, and for this reason the term 'eclipsing binary' is much more appropriate.

Algol is by no means unique. Many similar pairs are known, and the list runs into many hundreds; a few such stars, notably Lambda Tauri and Delta Libræ, are visible with the naked eye. Then there are the two extraordinary 'Kids', close to Capella, which have received a tremendous amount of attention from astronomers during the past decade or two.

The Hædi, or Kids, are three in number, and make up a

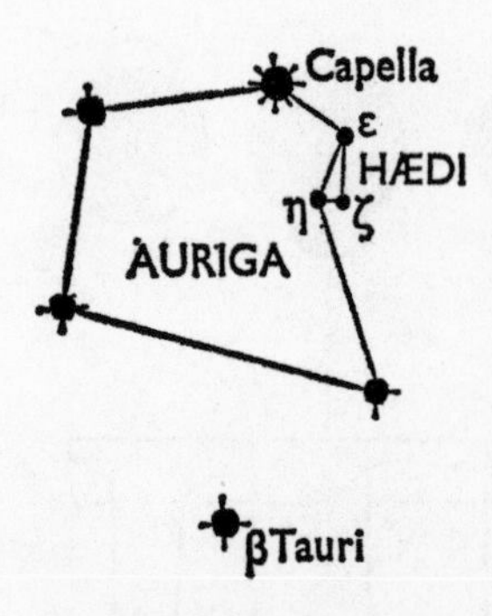

Fig. 50. The Hædi or Kids, near Capella. Zeta Aurigæ is a very interesting eclipsing binary; Epsilon Aurigæ is also a system of great importance—it may be an eclipsing binary, or it may include a black hole.

triangle which does not look superficially very exciting (Fig. 50). One of the trio, Eta Aurigæ, is entirely unremarkable. (For the record, it is a B-type star, about seventy times as luminous as the Sun.) The other two, Zeta and Epsilon Aurigæ, are eclipsing systems of very unusual kind.

Zeta—once dignified by a proper name, Sadatoni, which has fallen into virtual disuse—does not show violent changes in light, and the variations are not easy to follow with the naked eye. Also, the eclipses are infrequent, since the period is 972 days (as against less than three days for Algol). One component is a hot B7-type Main Sequence star, over 100 times as brilliant as the Sun, and with a diameter of some three million miles. The companion is a K4-type supergiant with a much lower surface temperature, but of immense size. Its diameter is over 200 million miles, larger than that of the Earth's orbit round the Sun (Fig. 51).

The most careful watch is kept upon Zeta Aurigæ when the red supergiant is eclipsing the B7 star. Even before the main eclipse begins, the blue star fades slightly, showing that it is shining through the vast, tenuous atmosphere of its companion. For some time after the start of the actual eclipse, the light from the B7 star is still to be seen, cutting through the outer layers of the supergiant. This shows that these outer layers must be highly rarefied, with a density of perhaps one five-millionth that of water. At this time the spectrum of the combined pair is very complicated, since we see both

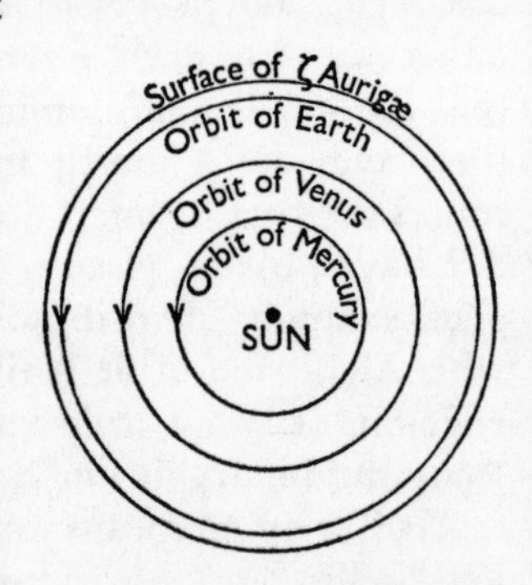

Fig. 51. Size of Zeta Aurigæ, as compared with the orbits of Mercury, Venus and the Earth round the Sun.

lines due to the supergiant and those of the blue star; but as the blue star passes behind the supergiant its spectrum is gradually blotted out, to reappear before the end of the eclipse.

Epsilon Aurigæ, the third of the Kids, is even more remarkable. It is certainly an eclipsing system with a period of 27 years; but as it may involve what is called a Black Hole it will be better for me to reserve it until Chapter 13.

Not all eclipsing systems are made up of giants or supergiants. Dwarf pairs exist as well, a good example being Castor C or YY Geminorum, which is an eclipsing binary made up of two Red Dwarfs. UX Ursæ Majoris, in the Great Bear, is different again. The components are not very unequal; each has about half the diameter of the Sun, but with a mass greater than that of the Sun, so that they are both over-dense and underluminous. The orbital period is only 4¾ hours, and each eclipse lasts for a mere forty minutes.

Another very famous eclipsing binary is Beta Lyræ or Sheliak. It is easy to find, since its position close to Vega makes it quite unmistakable, and its neighbour Gamma Lyræ makes an excellent comparison star. The variations in light were discovered in 1784 by Goodricke, only a short time before his death, and are quite unlike those of Algol. At maximum the magnitude is 3½, but there is no period of constant brightness, since changes are always going on. The sequence of events is always the same (Fig. 52). Starting from maximum, the star fades down to below the fourth magnitude; it then recovers, but when it fades again the minimum is less pronounced (magnitude 3¾). The succeeding minimum is again deep, so that deep and shallow minima take place alternately. The overall period is 12 days 22 hours 22 minutes—apparently increasing at a rate of about one minute in six years, which is slight but measurable.

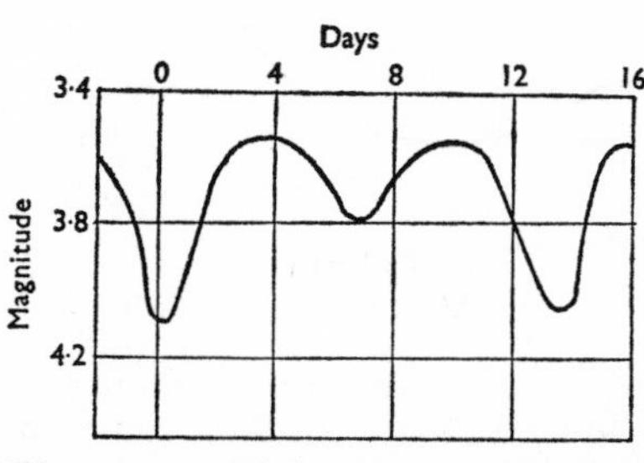

Fig. 52. Light-curve of the eclipsing binary Beta Lyræ.

The main component of Beta Lyræ is hot and bluish-white; it is also very powerful, and the distance has been given as

1,100 light-years. The secondary is less hot, but its spectrum has never been observed, so that nobody knows just what sort of star it is. The twins must be so close together that they almost touch, and each is drawn out into the shape of an egg; material is thrown off, and the stars must be surrounded by streamers of glowing gas. Only a few hundreds of Beta Lyræ-type stars are known, but there are also the W Ursæ Majoris variables, in which the components are equally close together and equally distorted by tidal forces, but are of later spectral type and considerably less luminous.

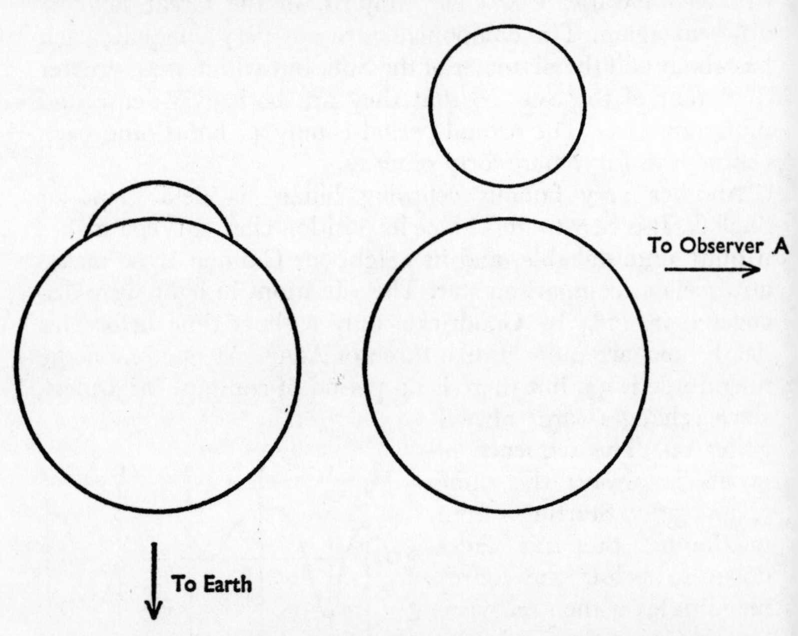

Fig. 53. Theory of an eclipsing binary. As seen from Earth, there will be eclipses; but if the same star were observed from position A, no eclipses would occur, and the star's light would remain steady.

I have included the eclipsing binaries in this chapter simply because they do show light-changes, though they are not truly variable. If we could look at Algol from some other vantage point, so that no eclipses would occur, the light would be quite steady, as shown in Fig. 53. So let us turn now to the genuine variables, beginning with a list of the various kinds—though let me say at once that any short list is bound to be open to criticism, because variable star studies are much more involved than used to be thought.

The main classes are:

1. Long-period variables, sometimes called Mira stars.
2. Semi-regular variables, such as Betelgeux in Orion.
3. Irregular variables, which are of various kinds.
4. Cepheids, with regular light-curves and short periods of from 3 and 50 days.
5. RR Lyræ stars, also quite regular, but with very short periods of less than a day.
6. Flare stars, which I have already described.
7. Novæ, or 'temporary stars'.
8. Supernovæ, involving the destruction of the star in its original form.

It seems fitting to begin with the 'Wonderful Star', Mira in Cetus (the Whale), which has the honour of being the first variable star to be recognized. The story began on 13 August, 1596, when David Fabricius, a Dutch pastor, was looking at Cetus and noted an object of the third magnitude. It seemed to be an ordinary star, but a few weeks later, when he looked again, it had disappeared. This was certainly odd, and one might have expected Fabricius to follow the matter up, but apparently he never did. (*En passant*, Fabricius met with a curious fate. In 1616 he preached a sermon in which he hinted that he knew the identity of the member of his congregation who had stolen one of his geese. Evidently his surmise was correct, because he was assassinated before he could divulge the name of the culprit.) Seven years after Fabricius' first observation, Johann Bayer was drawing up his famous catalogue when he saw the star again, and gave it the Greek letter Omicron. This time it was of the fourth magnitude, and Bayer did not connect his Omicron Ceti with the disappearing star of 1596. Again it vanished, and again nobody paid any attention. The star was seen now and again between 1603 and 1638, and eventually Phocylides Holwarda, also Dutch by birth, found that it appears and vanishes regularly.

Mira's period, or interval between successive maxima, is 331 days on average, but both the period and the magnitude

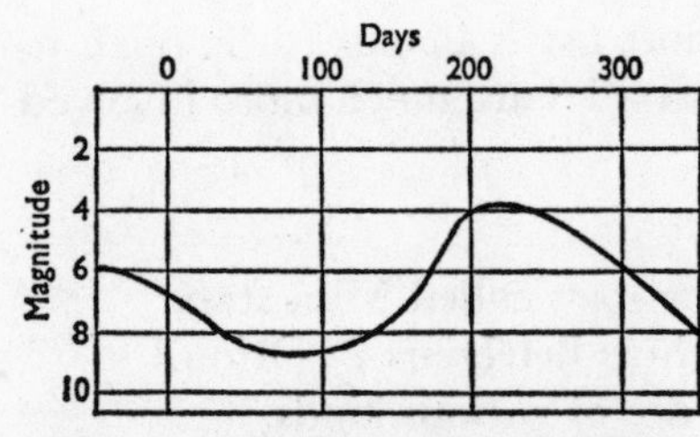

Fig. 54. Light-curve of Mira. The average period is 331 days.

range alter within reasonably narrow limits (Fig. 54). In August 1969, for instance, I saw Mira shining as a conspicuous star of the second magnitude—I made it 2·3 at its peak, only slightly inferior to the Pole Star—but at most maxima it does not even reach the third magnitude. At its faintest it falls to magnitude 10, so that it is beyond the range of ordinary binoculars. Actually, Mira is generally visible with the naked eye for only about 18 weeks of its 47-week period, so do not be surprised if you look up toward Cetus and fail to see it. (Of course, it is always easy to locate if you have an adequate telescope and a set of star-charts.)

Look at Mira, and you will notice that it is red. This is true of most of the long-period variables; their spectra are of type M or later. The diameter of Mira is believed to be of the order of 250 million miles, and its mass is ten times that of the Sun. Though Mira is over a hundred light-years away it is the closest of the long-period variables, and is the only one which can become really prominent with the naked eye.

Mira has a faint companion (which is either a White Dwarf or else some kind of sub-dwarf), but it is not an eclipsing binary; its variations are intrinsic. It alters both in size and in temperature. At its hottest the surface layer attains over 2,500 degrees, but at its coolest it is 700 degrees less. The star swells and shrinks, so that we must regard it as unstable. The spectrum, too, is variable, and at times there are bright lines as well as the familiar dark absorption bands always to be found in the spectra of Red Giants.

Over three thousand Mira-type variables are known in our Galaxy by now, but since they are all further a way than Mira itself they are not conspicuous. Even so, there are numbers of them within the range of the amateur observer equipped with a modest telescope. One is Chi Cygni, in the Cross of the Swan, which has an exceptionally large range. At its brightest it may reach the third magnitude, but at minimum it falls to below 14, which means that it disappears below the limit of even a fair-sized telescope. The average period is 407 days, and the spectrum is of the relatively rare type S, so that Chi Cygni is very red indeed. It is also a strong emitter in infra-red; if our eyes were sensitive to these long wavelengths, Chi Cygni would shine as one of the most brilliant objects in the whole sky.

Not many variables are bright enough to be given proper names or Greek letters, and a system was worked out which seems to be quite satisfactory. When a star was found to be variable, it was given the letter R in its particular constellation; the next variable was lettered S, and so on down to Z, after which the letters from R to Z were used in pairs. Then came the earlier letters of the alphabet. If all these were used up, the nomenclature scheme is to use a V followed by a suitable number; thus V334 Cygni was the first variable to be found in Cygnus after the end of the alphabetical series.

Among famous Mira-type stars are R Leonis in the Lion, near Regulus; R Cygni and U Cygni, in the Swan; R Leporis, the so-called 'Crimson Star' in the little constellation of the Hare; and U Orionis, near the borders of Orion and Taurus. All are easy to see with binoculars when at maximum, but all become faint when near minimum. U Orionis is an awkward star. Its period is 372 days, so that it comes to maximum only about a week later every year; and at present (1974) maximum occurs in the summer, when the star is too near the Sun to be observed at all.

All the Mira stars are somewhat erratic in both range and period, but in one case, R Hydræ, there seems to have been a systematic change. The period was about 500 days when the fluctuations were first noted, by Maraldi in 1704, but nowadays the period is less than 400 days. The change seems to be real, even though it would be dangerous to put too much faith in the old observations. Whether it represents something permanent, connected with the star's evolution, we do not know.

Next we come to the semi-regular variables, which behave in a rather different way. Their fluctuations are often very slow, and they do not generally have a range of more than a couple of magnitudes, against over ten magnitudes for the Mira-type star Chi Cygni. The periods of the semi-regular variables are very rough, and one can never predict quite what they will do next. Betelgeux in Orion is a classic example.

Officially the period is given as 2,070 days, or between five and six years, but the irregularities are so marked that one can hardly give it any true period at all. When at its best it can almost match Rigel; at its faintest, it is comparable with Aldebaran. Here too we have a swelling and shrinking process, together with a change of temperature which is, admittedly, fairly slight.

Vast though Betelgeux may be, the semi-regular variable Rasalgethi or Alpha Herculis is even larger. Its diameter may be over 400 million miles, and it is surrounded by an extensive envelope of thin gases spreading out to over 600 million miles, so that it is a supergiant of extreme type. Even though it is over 400 light-years away it is quite bright; it ranges between magnitudes 3 and 4, and is always easy to see with the naked eye. It is not hard to identify, because the nearby second-magnitude star Rasalhague or Alpha Ophiuchi acts as a marker for it.

Rasalgethi may give us some clue as to the real nature of these semi-regular stars. In the outermost layers, the gaseous envelope may condense into patchy clouds, which could do something to veil the red star and cause a fall in apparent brightness, though undoubtedly the star itself is intrinsically variable as well.

Passing over the so-called RV Tauri stars, which are G to K type giants and have reasonably well-marked periods with alternate deep and shallow minima, and doing no more than mention the variables with a very small magnitude range (too small to be detected without sensitive equipment) we come to the eruptive stars, which again are of various kinds. Among the most interesting are the 'dwarf novæ', otherwise called U Geminorum or SS Cygni stars, since U Geminorum was the first of the class to be discovered (by the English astronomer J. R. Hind, in 1855) and SS Cygni is the brightest and most closely studied of them (Fig. 55).

A dwarf nova stays at its minimum brightness for most of the time, with only slight fluctuations. At fairly predictable intervals there is an outburst, and the star brightens up by several magnitudes, remaining at maximum for a few days before declining once more. Not all maxima are of the same length or the same amplitude, and neither are the intervals

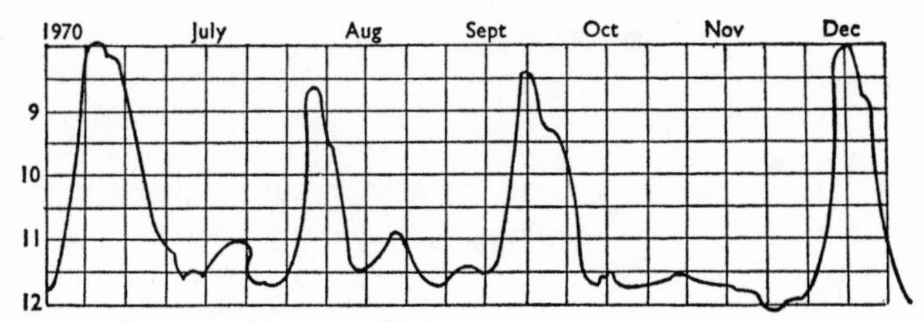

Fig. 55. Light-curve of SS Cygni in 1970, from observations made with my 12½in reflector.

between successive maxima constant. With SS Cygni the average interval is fifty days, but with U Geminorum it is 103 days, and with the much fainter star UV Persei it is as much as 300 days.

For some time nobody was sure whether these strange variables were very luminous and remote, or relatively nearby and feeble. It now seems that they lie within a few hundred light-years of us, and are dwarfs. It has been found, too, that they are close binaries; a typical system is made up of a rather underluminous yellow star with a surface temperature about the same as that of the Sun, together with a White Dwarf. The exact nature of the outbursts is still uncertain. It is not likely that they originate in the yellow star; they could be due to the White Dwarf, but are more probably bound up with streams and clouds of gas associated with both components. There have been suggestions, too, that the SS Cygni stars evolve from the W Ursæ Majoris type eclipsing binaries, in which, as we have seen, we have two yellow dwarfs orbiting each other in virtual contact, so that the movements of the associated gas-clouds and streams are very complicated. Variable star observers are also intrigued by the Z Camelopardalis stars, which are not too unlike the dwarf novæ except that they sometimes seem to forget that they are variable, and stay almost constant for weeks or months on end before resuming what to them is normal behaviour.

The next irregular variables are quite different. The best-known member of the class is R Coronæ, in the Northern Crown, not far from Arcturus in the sky (Fig. 56). Less than fifty R Coronæ stars are known, and most of them are always very faint. R Coronæ itself is the exception; normally it is of about the

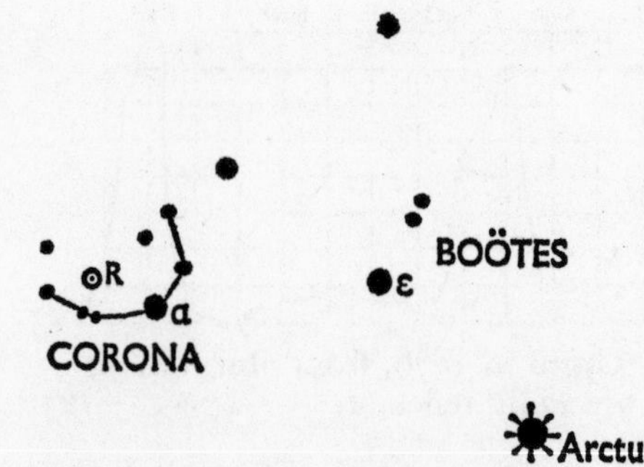

Fig. 56. Position of the remarkable variable star R Coronæ.

sixth magnitude, on the fringe of naked-eye visibility, and changes very little, but at unpredictable intervals it falls to a deep minimum, sometimes becoming so dim that small telescopes lose it completely (Fig. 57). For instance, it went through minimum in April 1972, and dropped to below the twelfth magnitude; it had begun to decline in late February, and not until the late summer was it back to normal. On the other hand there were no minima at all between 1925 and 1934. This was before my time, but I imagine that observers began to lose patience with its reluctance to 'perform'!

R Coronæ stars are very poor in hydrogen, but are rich in carbon. It has been suggested that the minima may be due to the building-up of carbon particles in the very outer part of the star's atmosphere, resulting in blanketing of the radiation, so

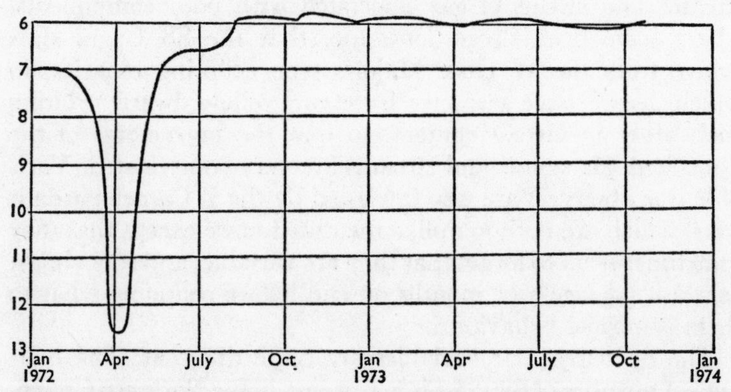

Fig. 57. Variations of R Coronæ, 1972–4, from my own observations made at Selsey. There was one principal minimum. Otherwise the star remained almost constant at the sixth magnitude.

that in fact the star temporarily hides itself coyly behind a veil of soot. The spectral changes are very complex indeed, and nobody can pretend that we really know how R Coronæ stars function. They are giants of high luminosity, and are probably well advanced in their life-stories. It is a pity that there are so few of them.

Though R Coronæ stars are old, irregular variables of the T Tauri type are certainly very young. Indeed, they are still contracting toward the Main Sequence, and many of them are involved in nebulosity. Incidentally, stars of this kind are very strong infra-red emitters, and theorists are becoming more and more interested in them. Their fluctuations seem to be quite random, so that clearly they have yet to settle down to their period of steady, sober existence.

As a slight but justifiable digression, this seems to me the moment to return to two other remarkable stars which are undoubtedly very young, and must be classed as variable inasmuch as they show undoubted changes in output. In 1936 a very faint star in Orion, now called FU Orionis, suddenly increased to 250 times its usual brightness, and after five months it had reached the tenth magnitude, since when it has stayed almost constant. In 1969 another star, this time in Cygnus (V.1057 Cygni: the constellation contains so many variables that the alphabetical series was exhausted long ago) brightened up by six magnitudes, taking 250 days to do so. It also has remained almost constant ever since. It seems that these two stars, both of which are involved in nebulosity, abruptly 'blew away' their surrounding dust-clouds, so that for the first time we could see them properly; they too are very strong in the infra-red, and it is logical to associate this long-wavelength radiation with heated dusty material. Over the years, no doubt other youthful stars will be seen to burst forth in the same way.

But of all variable stars, the most important to astronomers are the Cepheids. They take their name from Delta Cephei, in the far north, whose fluctuations were discovered in 1784 by the keen-eyed Goodricke. The chart shows how to find it (Fig. 58); two members of the W of Cassiopeia show the way to it, and it is easily located, because it forms a small triangle with its neighbours Epsilon and Zeta Cephei. The magnitude ranges from 3·7 to 4·3 in a period of 5 days 9 hours, and the period

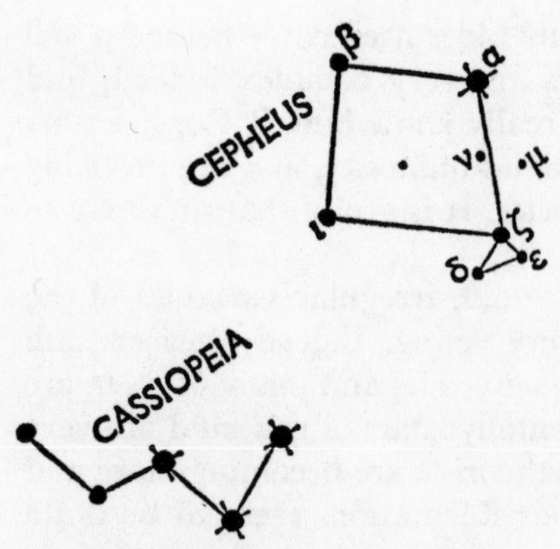

Fig. 58. Position of Delta Cephei. The red irregular variable Mu Cephei is also shown; although Cepheus is not a bright or well-marked group it is easy to locate by using the W of Cassiopeia as a guide. It is, of course, circumpolar all over Britain.

is absolutely regular, so that it may be worked out to within an accuracy of one second. The light-curve (Fig. 59) is not completely smooth, as the increase from minimum to maximum is steeper than the subsequent drop, but we can always tell how bright Delta Cephei will be at any particular moment.

This time we are dealing with a regular pulsation, involving changes in diameter, temperature and spectrum. The temperature range amounts to as much as 2,000 degrees, and the spectral type changes between F4 and G6. Like all its kind, Delta Cephei is a yellow supergiant. Other naked-eye Cepheids are known; in the northern hemisphere there are Eta Aquilæ and Zeta Geminorum, and in the southern sky Kappa Pavonis and Beta Doradûs. Telescopic Cepheids are numerous, and over six hundred have been found in our Galaxy alone. Because they are powerful stars, they are visible over tremendous distances.

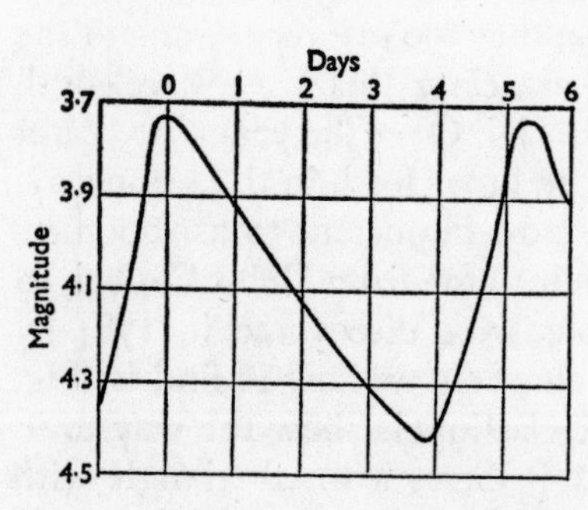

Fig. 59. Light-curve of the variable star Delta Cephei: period 5·3 days.

Cepheids are of various classes. Delta Cephei itself is the prototype of the classical or type I Cepheid; type II Cepheids, less common and less luminous, are sometimes called W Virginis stars after their best-known member; and we have Cepheid-type stars of very small magnitude range. One of these is none other than the Pole Star. The period is just under four days, but the changes in brightness are too small to be noticed with the naked eye.

Cepheids are pulsating stars, all of which have evolved off the Main Sequence. In a star there are two forces to be considered: the weight of the material, tending to make the star shrink, and the pressure of the hot gases within. In a stable star these forces balance each other out, but in an unstable star such as a Cepheid they do not. If we begin the cycle with a contraction, the pressure inside will build up until it has become great enough to overcome the effects of the weight of the outer layers, and the shrinking will stop, to be followed by expansion. Now the weight of the overlying layers will decrease, but the pressure inside will decrease faster—and so after a while contraction will begin again. Once more there will be an 'overshoot', and so the pulsations continue. The effect may be likened to that of a spring, though the analogy is not really very accurate.

The pulsation time for a very large, rarefied star would logically be expected to be longer than that for a smaller, denser star, and it would not be surprising to find that the period of variation of a Cepheid would be linked with its real luminosity. This is in fact the case, and the 'period-luminosity' law is well-defined. The rule is: The longer the period, the more powerful the star. Thus Zeta Geminorum (period 10·1 days) is more luminous than Delta Cephei itself (period 5·4 days). And since the two stars appear almost exactly equal in brightness as seen from the Earth, Zeta Geminorum must therefore be the more distant of the two.

The original discovery was more or less fortuitous. In 1912 Miss Henrietta Leavitt, at Harvard, was studying photographs of one of the southern star-systems, the Small Magellanic Cloud, about which I will have more to say later. The Cloud lies well beyond our Galaxy (though admittedly this was not known in 1912). It contains plenty of Cepheids, and in studying them Miss Leavitt found that the apparently brighter variables always had the longer periods.

The essential point here is that to all intents and purposes the variables in the Small Magellanic Cloud can be regarded as being at the same distance from us—just as we can say that for all practical purposes Chichester and Bognor Regis are the same distance from New York. Therefore it followed that the brighter Cepheids were genuinely the more luminous, and within a few years a definite relationship had been worked out.

This would have been important in itself, but it proved to be even more useful than had been thought. A Cepheid gives away its luminosity merely by the way in which is behaves. And if its real brightness and its apparent brightness are both known, its distance can be calculated. The best analogy I can give is that of a man on a sea-shore, looking at a light out across the water. If the light appears dim, it can be either a genuinely dim lamp close at hand, or else a powerful light a long way away. If our observer knows how strong the light really is, he can get a good idea of its distance.

Cepheids, then, provide invaluable 'measuring beacons', and because they are luminous enough to be seen in galaxies far beyond our own we depend upon them to a great extent. We can also make use of the shorter-period pulsating stars, known as RR Lyræ variables, all of which have periods of less than a day, and all of which seem to be of approximately the same luminosity—rather less than 100 times that of the Sun.

Professional astronomers pay great attention to the Cepheids and the RR Lyræ stars, as well as to the eclipsing binaries and the flare stars. With sensitive equipment (mainly photo-electric) used together with large telescopes, the magnitudes can be measured with great accuracy, and of course the various spectral changes are under constant survey. So far as the long-period and most of the irregular variables are concerned, the need for absolute precision is not so great. It does not matter very much whether the magnitude of, say, U Orionis is 9·5, 9·6 or 9·7 at any particular time; an accuracy of a tenth of a magnitude is good enough for most purposes, and this is where the amateur comes in.

With the development of space-probes to the Moon and planets, it would be idle to claim that the rôle of the amateur so far as the Solar System is concerned is as extensive as it used to be. For example, mapping the Moon is now virtually complete, and it is strange to reflect that only a couple of decades ago the best lunar charts were amateur ones! Bearing this in mind, the amateur has looked around for new fields, and variable star work is becoming more and more popular. This is particularly the case because a modest telescope can be very useful, and even binoculars can be pressed into service.

The average amateur makes his estimates by using comparison stars which do not alter. To show what is meant, let me give an example from my own observation book. Just before writing this chapter, I was using the 12½-inch reflector at my observatory in Selsey to make an estimate of W Lyræ, a typical Mira-type star with a period of around 196 days and a magnitude range of from 7½ to 13. I used two comparison stars, lettered *d* and *c* on my charts (Fig. 60), whose magnitudes I knew; *d* is listed as 9·4 and *c* as 9·0. I made W Lyræ 0·3 magnitude brighter than *d*, and 0·1 fainter than *c*. Clearly, then, the magnitude of the variable was 9·1. I am not prepared to say that I could not have made a slight error, but I would be very surprised to find that I could be very far wrong in my estimate.

There is, therefore, a tremendous amount of scope. I hope

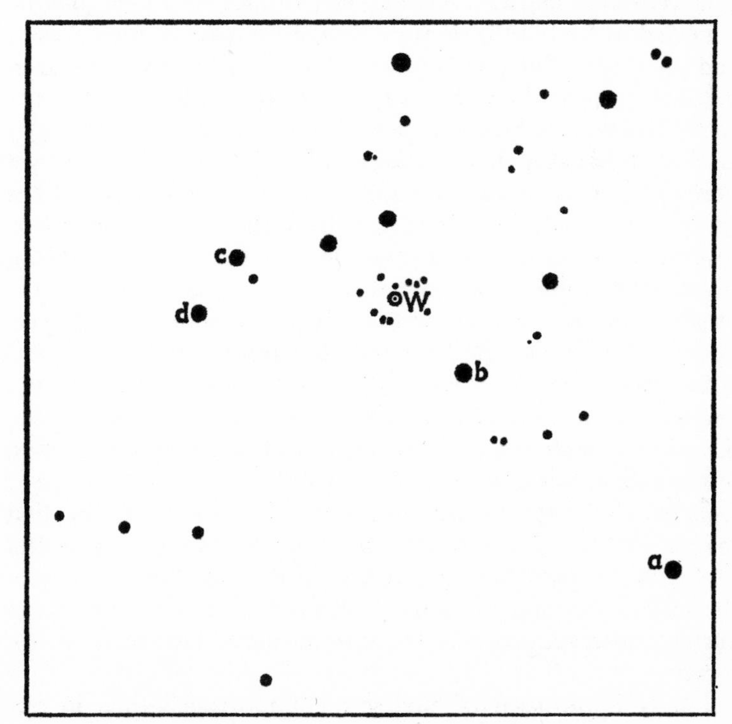

Fig. 60. Position of the 'binocular variable' W Lyræ.

that at least some readers will feel inclined to take up the work, because it is useful as well as enjoyable—particularly, I think, with those variables which are always apt to spring surprises on us.

Can the Sun be regarded as a variable star? Strictly speaking, no; its output of light and heat remains virtually constant, which is very lucky for us, since even a small alteration would result in our being fried or frozen. Yet we know that there is a semi-regular cycle of about eleven years, and that the numbers of sunspots wax and wane in a marked manner. Sunspots are associated with magnetic fields, and these follow a definite pattern which also is linked with the solar cycle. Therefore we may regard the Sun as magnetically variable, even though its overall field is weak.

Finally, I must say something about the 'secular variables', where the changes are slow enough to be imperceptible except over very long periods. The first star-catalogues which may be regarded as fairly reliable were drawn up many centuries ago, and there are a few cases of stars which seem to have altered in brightness since then. The best example is Megrez or Delta Ursæ Majoris, in the Great Bear. Its present magnitude is 3·3, so that it is more than a magnitude fainter than Mizar; yet Ptolemy and the other old observers made it equal to the other stars of the Bear, so that it may have faded. Even now it is suspected of occasional fluctuations, so that it could be an irregular variable which remains steady for many years at a time. Castor in Gemini used to be ranked superior to its 'twin', Pollux, but is now half a magnitude fainter, so that if the old estimates are correct either Castor has decreased or Pollux has brightened up. Acamar or Theta Eridani, the 'Last of the River', was included in Ptolemy's list of first-magnitude stars, but is now below the third (it is, incidentally, a fine binary); Denebola or Beta Leonis may have dropped from the first magnitude to below the second, while Rasalhague or Alpha Ophiuchi, formerly ranked of the third magnitude, is now of the second. We must beware of placing too much faith in the ancient observations, but the discrepancies deserve to go on record.

I may be accused of having spent too much time in discussing variable stars, bearing in mind that I am trying to do no more than write a basic book. This may be true; but I plead justification simply because so many amateurs now concentrate upon variable star work. And these stars are fascinating even to non-theorists. They are of so many kinds: the 'fake' variables such as Algol, which turn out to be nothing more than eclipsing binaries; the long-period Mira stars; the punctual and highly-reliable Cepheids, and the erratic rebels which we can never predict. Ultimately, theory depends upon observational work, and there is plenty to be done. Much can be learned from studying these strange, unstable suns so many light-years away from us.

Chapter Eleven

EXPLODING STARS

THERE WAS A TIME, not so many centuries ago, when the starry sky was regarded as changeless. True, the Moon and planets wander about; but their movements are predictable, and the stars themselves were thought to show no alterations at all, either in position or in brightness.

It is quite correct to say that the constellation patterns are to all intents and purposes unchanging, and that most of the stars shine steadily in light, but now and then what seems to be a new star bursts forth in a position where no conspicuous star has been seen before. Such was the case on 13 December, 1934, when an English amateur astronomer, J. P. M. Prentice, was taking a nocturnal stroll after a spell of meteor-watching and saw something very peculiar about the area between Draco and Hercules. A new star of about magnitude $3\frac{1}{2}$ had appeared, and Prentice realized at once that it must be what is termed a nova. For the next few months it remained visible with the naked eye, though it subsequently faded away and has now become very faint (Fig. 61).

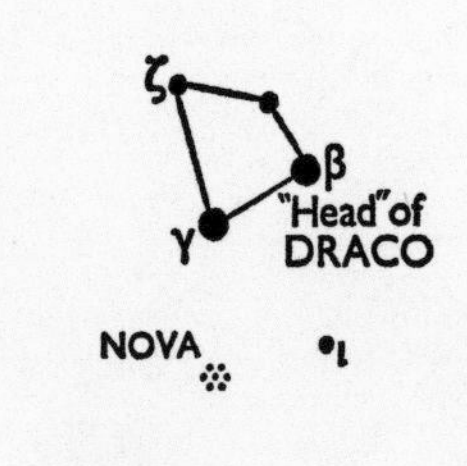

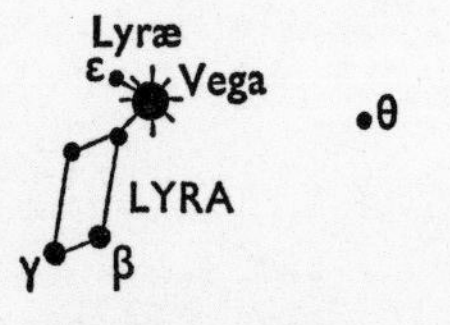

Fig. 61. Position of Prentice's nova of 1934 (DQ Herculis).

Other novæ have been seen, both before and since. For example, another English amateur, G. E. D. Alcock—by profession a schoolmaster, teaching in Peterborough—discovered a particularly interesting nova in the little constellation of Delphinus (Fig. 62), in July 1967. This time the magnitude was about $5\frac{1}{2}$, so that the nova was not prominent with the naked eye; but Alcock had been carrying out a systematic search, using wide-field, specially-mounted binoculars, and had spent years in memorizing the position and magnitudes of over thirty thousand stars, so that he could recognize a newcomer on sight. At once he contacted other observers, and the discovery was confirmed. I was one of those notified, and I was able to identify the nova without any trouble at all, though I would be the last to claim that I would have noticed it unless George Alcock had told me where to look!

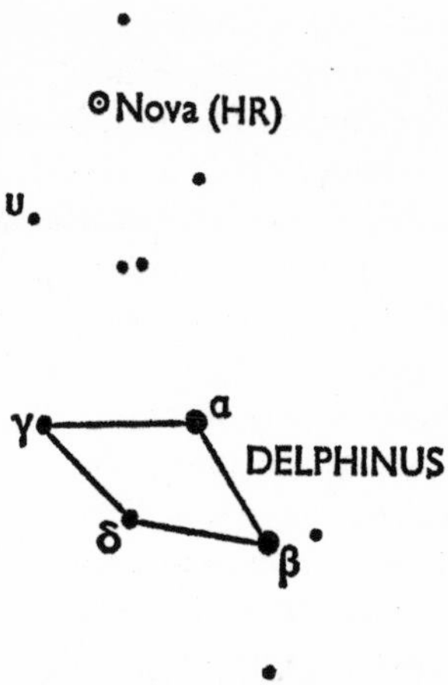

Fig. 62. Position of Alcock's nova, HR Delphini of 1967. At its peak it was easily visible to the naked eye, but by the beginning of 1974 it had dropped to the eleventh magnitude.

The word 'nova' is Latin for 'new', but to be accurate a nova is not a new star at all. What happens is that a formerly faint star suffers a tremendous outburst below its outermost layers. A surface 'shell' is ejected at a speed which may be as much as three or four hundred miles per second, and the result is a striking, though temporary, increase in apparent luminosity, even though the material sent hurtling outward cannot exceed about 1/10,000 of the total mass of the star. Sometimes the shell becomes visible, as happened with a brilliant nova which flared up in 1918 in Aquila. About four months after the outburst the gas-cloud was first seen, and it remained on view for twenty years, though it slowly spread out and became fainter. By the time that it dropped below the threshold of visibility its diameter had reached something like a million million miles.

Naked-eye novæ are less uncommon than might be thought, and no less than twenty-two have been seen so far during the present century. The year 1936 was remarkable inasmuch as it produced three—another nova in Aquila, which reached a maximum magnitude of 5·4; one in Lacerta, rising to 1·9, so that for a brief period it outshone the Pole Star; and a third in Sagittarius, where the peak magnitude was 4·5. Amateurs, who know the skies so well, have a fine record in nova-detection, and Alcock has now found three. His first success was with Nova Delphini in 1967; he followed this up with Nova Vulpeculæ in

1968, and finally a much fainter star in the little constellation of Scutum, the Shield. (As he has also discovered four comets, one cannot but admire his patience and skill.)

Obviously it is highly desirable to study the spectrum of a nova as soon as possible after its outburst, because the activity is then at its height, and tremendous disturbances are taking place. It is not nearly so easy to study the pre-nova condition, because we can never tell which stars are likely to explode, and there is no warning at all. Still, in some cases pre-outburst spectra are available, and it seems that stars liable to nova outbursts are generally hot sub-dwarfs of spectral type A or thereabouts. There is a strong likelihood that a nova is a star which is coming to the end of its stable career, and is on the way toward collapsing into the White Dwarf condition; moreover, it has been found that many novæ are close binaries—sometimes, as with Prentice's star of 1934 (DQ Herculis) eclipsing binaries. On the other hand, there are difficulties in the way of supposing that every normal star becomes a nova at some time during its evolutionary career. If this were so, we might expect to see far more novæ than we actually do.

Originally it was thought that a nova might be the result of a head-on collision between two stars, but this attractive idea has been definitely ruled out of court. The stars are so widely separated in space that collisions can hardly ever occur, and novæ are much too common to be explained in any such way.

If a nova lies in a region where there is a good deal of thinly-spread gas and dust, it will light up the material. This happened with the brilliant nova seen in Perseus in 1901, and which became as bright as Capella. At first it was thought that the gas-cloud round the star must be expanding rapidly, but this was not really so; the radiation from the nova was spreading out in all directions and illuminating more and more of the gas and dust, so that the cloud gave the false impression of expanding at the speed of light itself. The cloud is shown in Plate II.

Not all novæ are alike. Each has its own characteristics, so let us consider two, which were very different in their behaviour. On 7 June, 1918, a newcomer appeared in Aquila, and by the following evening it had reached the first magnitude. For a short time on 9 June it attained magnitude −1·4, the same as that of Sirius, so that it was 75,000 times as brilliant as it had been only three days earlier. It did not stay bright for long; by the end of June it was below the third magnitude, and then faded steadily, until by late 1924 it had declined to about magnitude 11. Examination of earlier photographs of the region showed that it had also been of magnitude 11 before its outburst, so that its spell of glory was brief (Fig. 63).

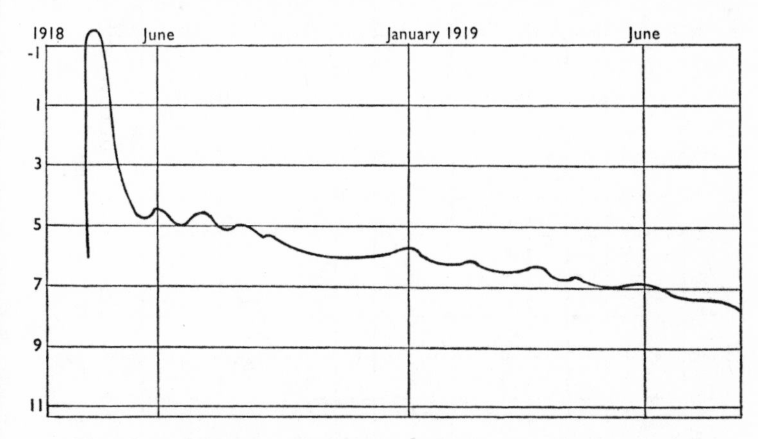

Fig. 63. Light-curve of Nova Aquilæ, 1918—a fast nova.

Incidentally, it would be wrong to suppose that the explosion really happened in 1918. The star's distance is of the order of a thousand light-years, in which case the flare-up dates back well before the Norman Conquest; we were looking at an event which had taken place in the remote past. Before it turned into a nova, the star was rather more powerful than the Sun; at its peak it gave out at least a million times as much energy as the Sun.

Nova Aquilæ blazed up before I was born, and I have seen it only in its modern, obscure form, but I have been able to follow the second nova to be described here: HR Delphini (Fig. 64). This was the star discovered by Alcock on 8 July, 1967. Unlike Nova Aquilæ, it did not become glaringly brilliant, but neither did it fade so quickly. It fluctuated between magnitudes 5 and 4 until around the end of the year, and in December it reached its greatest brightness—about magnitude $3\frac{3}{4}$. Thereafter it faded; the magnitude was still 5 on 1 March, 1968, 6 in

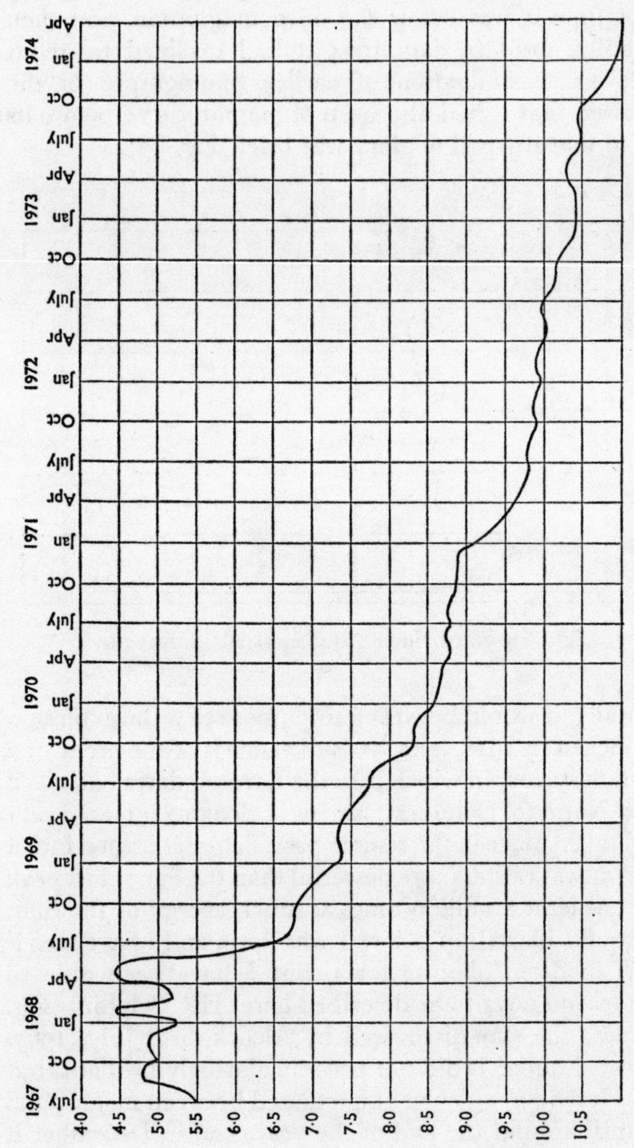

Fig. 64. Light-curve of Nova (HR) Delphini, from its discovery in 1967 up to 1974, from my own observations.

July, 7 in October and, after a further slowing-down in the decline, about 8 in August. Not until early 1971 did it drop to the ninth magnitude, and at the end of 1973 it was still above magnitude 11. The light-curve given here, drawn from my own estimates, shows how different was its behaviour from that of the spectacular but shorter-lived Nova Aquilæ. Remember, though, that HR Delphini is much the more remote of the two. Its distance seems to be about 30,000 light-years, so that the outburst took place in prehistoric times even though we have only just had the news of it.

It is not likely that HR Delphini will ever disappear. Its pre-nova magnitude was 12, and it was then a very hot star of type O. Probably it will return eventually to its original condition, apparently none the worse for its outburst.

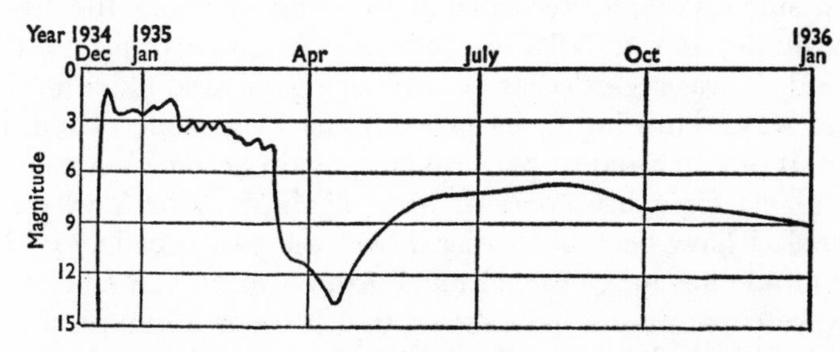

Fig. 65. Light-curve of DQ Herculis, the 1934 nova: a slow nova.

The first bright nova which I was able to observe was the famous DQ Herculis of 1934. The light-curve was exceptional; after a rather long maximum there was a sudden decrease, followed by a gentle rise before the final fading began (Fig. 65). Today, as I have said, it is an eclipsing system; the period is only 4 hours 39 minutes, and in addition the old nova itself seems to be fluctuating quickly over a small magnitude range. The spectrum is made up of a continuous background, crossed by emission lines which are produced partly in the large expanding nebulous envelope blown off during the 1934 outburst and partly in a much more compact nebula surrounding the old nova. The companion star has never been seen, but is believed to be an M-type dwarf, while the old nova is probably so dense that it is not unlike a White Dwarf. It is a tremendous pity that we have no record of the spectrum of DQ Herculis before its flare-up. During its outburst it threw off no less than eleven shells, and at one period, when it was declining, it shone with a decidedly greenish hue. The distance is about 800 light-years, making it the closest of modern bright novæ. Another first-magnitude nova, seen in Puppis in 1942, was more remote and much more luminous; at its peak the luminosity was around 1,600,000 times that of the Sun.

There has been much discussion as to whether a nova is to be regarded as 'something apart', or merely as a peculiar kind of variable star. We even have a few cases of stars which have burst forth more than once. One of these is T Coronæ in the Northern Crown, nicknamed the Blaze Star. In 1855, Argelander, a German astronomer who compiled a particularly good star-catalogue, recorded it as being of about the tenth magnitude, but in 1866 it suddenly brightened up to the second. It was regarded as an ordinary nova, and faded in the usual way, returning to its pre-outburst magnitude. Then, in 1946, it put on a repeat performance, again becoming a naked-eye object for a brief period. Now, again, it is back where it started; I have been observing it over the past decade, and its magnitude has never varied much from a mean value of 10·0. Spectroscopic studies have shown that it too is a binary, made up of a highly luminous B-type star, responsible for the nova outbursts, and a Red Giant which is itself irregularly variable over a small range.

If T Coronæ blazes out every eighty years or so, we must pay close attention to it around A.D. 2026, though by then I will have reached the advanced age of a hundred and three and will probably be unable to do it justice! Another 'recurrent nova', RS Ophiuchi, has shown maxima in 1901, 1933, 1958 and 1967, and there are a few more stars of the same class.

P Cygni, in the cross of the Swan, is different again, and is not easy to classify. It was recorded as a third-magnitude star in 1600, and then slowly faded away until it had dropped below naked-eye visibility. In 1654 it reappeared, and after minor fluctuations settled down to the fifth magnitude in 1715, where it has remained ever since. Spectral studies show that mass is constantly streaming out from its surface, though not with the tremendous velocities encountered in nova explosions;

it is clearly unstable, and is also very luminous. Other P Cygni type stars are known, though they are uncommon.

Even odder is Eta Carinæ, in the Keel of the Ship. Nowadays it cannot be seen with the naked eye, but it has certainly known glory. Before 1820 it was reasonably conspicuous, and in 1837 and early 1838 it shone more brilliantly than any star in the sky apart from Sirius; it was still of the first magnitude in 1856, and it did not vanish from naked-eye view until 1867. In every way Eta Carinæ is an exceptional object, and northern observers never cease to bemoan the fact that it never rises above the European horizon.

Though P Cygni, Eta Carinæ and other erratic stars cannot be classed with the novæ, we must remember that irregular variables such as SS Cygni and U Geminorum show smaller-scale, much more frequent outbursts, so that there may be some sort of a link with the recurrent novæ at least—though SS Cygni stars are dwarfs, and the association may not be as close as might be thought, in which case the common term of 'dwarf novæ' for them may be rather misleading.

If our Sun became a nova, life on Earth would be destroyed at once. Fortunately, nothing of the kind is likely to happen; the Sun is steady and reliable, and it is not the sort of star to suffer an outburst—though what will happen in the far future, thousands of millions of years hence, we cannot tell.

Generally speaking, everything in the universe tends to happen with majestic slowness; but this does not apply to the novæ. Developments take place so rapidly that we find it hard to keep pace with them as a formerly obscure star bursts forth, flaring up into a true celestial searchlight, its spectrum changing constantly as it pours energy and material into space. It is a reminder to us that the sky is not static, as the ancients believed, and that at times we can watch phenomena infinitely more dramatic than anything which can ever happen on our tiny Earth.

Chapter Twelve

SUPERNOVÆ—AND PULSARS

IN THE YEAR 1054, Chinese and Japanese astronomers were surprised to see a brilliant new star in the constellation of Taurus, the Bull. It was so bright that it could be seen even with the Sun above the horizon, and it lasted for several months before it faded away. No accurate observations of it were made, and after a while the strange 'guest star' was more or less forgotten.

Then, in 1572, there came another outburst of the same kind, in another part of the sky. On 11 November the young Danish stargazer Tycho Brahe, later to become the greatest observer of his time, made a discovery which was to alter not only his life, but also—to some extent, at least—the whole history of astronomy. In his own words:

"In the evening, after sunset, when, according to my habit, I was contemplating the stars in a clear sky, I noticed that a new and unusual star, surpassing all the other stars in brilliancy, was shining almost directly above my head; and since I had, almost from boyhood, known all the stars of the heavens perfectly (there is no great difficulty in attaining that knowledge), it was quite evident to me that there had never before been any star in that place in the sky, not even the smallest, to say nothing of a star so conspicuously bright as this. I was so astonished at this sight that I was not ashamed to doubt the trustworthiness of my own eyes. But when I observed that others, too, having this place pointed out to them, could see that there really was a star there, I had no further doubts. A miracle indeed, either the greatest of all that have occurred in the whole range of nature since the beginning of the world, or one certainly that is to be classed with those attested by the Holy Oracles."

Tycho was an extraordinary man. He was proud, grasping and frequently cruel; he was also a firm believer in astrology, the superstition of the stars; and he could never bring himself to believe that the Earth could be a planet moving round the Sun. To him, the Earth was the supreme body in the universe.

As the months passed by, and the new star faded, Tycho did not miss the opportunity to point out that the effects on humanity would be dire:

"The star was at first like Venus and Jupiter, giving pleasing effects; but as it then became like Mars, there will next come a period of wars, seditions, captivity and death of princes, and destruction of cities, together with dryness and fiery meteors in the air, pestilence, and venomous snakes. Lastly, the star became like Saturn, and there will finally come a time of want, death, imprisonment and all sorts of sad things."

Despite his brilliance as an observer, Tycho had no idea of the nature of the star. Neither could he follow it as soon as it sank below the sixth magnitude; this was before the age of telescopes. Then, in 1604—three years after Tycho's death—another spectacular new star burst forth, this time in the constellation of Ophiuchus. Johannes Kepler, later to use Tycho's observations to disprove Tycho's theory of the universe (a strange twist of fate!) was among those who watched it, and it is usually called Kepler's Star. It too became, probably, as bright as Venus; then it too faded away and was lost to view.

Since then we have seen no more new stars as striking as these. They were much more dramatic than the ordinary novæ discussed in the last chapter. They were in fact supernovæ, or stellar deaths. At its peak, a supernovæ of the most violent type may shine with a luminosity equal to that of around two hundred million Suns, though clearly it cannot maintain this fantastic output for long.

Later searches of the old records showed that a star seen in the constellation of Lupus, the Wolf, in 1006 was also a supernova, but the Chinese 'guest star' of 1054 is the most important of all so far as modern astronomy is concerned, because we can still see its débris. Near the third-magnitude star Zeta Tauri, not far from the border of Orion, even a small telescope will show a dim, misty patch. Photographs taken with giant telescopes show an amazingly complex structure, with gaseous filaments; we can make measurements to prove that the whole gas-cloud is expanding, and we know that in addition to sending us visible light it is a source of radio waves, X-rays and gamma-rays. Officially it is called M.1, because it was the first entry in Charles Messier's famous catalogue of nebulæ and star-clusters, about which more anon. Lord Rosse, the great last-century Irish astronomer, nicknamed it the Crab Nebula because of its shape, and this is what we always call it today. There is absolutely no doubt that the Crab is the wreck of the 1054 supernova. The gas is expanding outward at the rate of over 600 miles per second, and if we calculate backward in time we find that the expansion must have begun about A.D. 1054.

Of course, our information is out of date—as always happens when we study any object well beyond the Solar System. The distance of the Crab is some 6,000 light-years, and this puts the true date of the outburst at 5000 B.C., before there were any Earthly astronomers to watch it. Suppose that there are alien scientists living on a planet moving round a star 1,000 light-years from us, 'behind' us with respect to the Crab? They will not yet have seen the outburst at all; but if they are on the alert, they will do so within the next century, as soon as the light-waves reach them.

It has been said that there are two branches of modern astronomy: the astronomy of the Crab Nebula, and the astronomy of everything else! This may be something of an exaggeration, but it would be hard to over-emphasize the importance of the Crab from the theoretical point of view. To begin with, it contains one of the remarkable objects known as pulsars, and it is one of the strongest radio sources in the sky. Before going any further I must, I think, digress temporarily in order to say something about the subject of radio astronomy, because it is so vital to the whole problem of supernovæ.

Light may be regarded as a wave-motion, and its colour depends on its wavelength—that is to say, the distance between one wave-crest and the next. You can easily see what is meant, simply by throwing a stone into a calm pond and noting the ripples; but with light, the wavelengths are so short that to measure them in inches or even millimetres would be cumbersome. Instead, we use the Ångström, named after the last-century Swedish physicist Anders Ångström. One Ångström is equal to a hundred-millionth of a centimetre. The line in the diagram is one centimetre long (Fig. 66). If you mentally

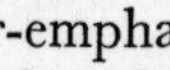

Fig. 66. A centimetre line. From this, you can judge how tiny an Ångström unit really is.

try to divide it into a hundred million parts, you will appreciate how small an Ångström unit really is.

The shortest light which affects our eyes is violet, with a wavelength of the order of 4,000 Å. If the wavelength is less, the 'light' is invisible: that is to say, it is ultra-violet. With still shorter wavelengths we come to X-rays, and then to gamma-rays. The Crab sends us all these.

Red light, at about 7,600 Å, lies at the long-wavelength end of the visible range. With increased wavelength we come first to the infra-red (most people are familiar with the infra-red lamps used in hospitals) and then to what are known as radio waves, where the wavelengths may amount to several miles. Let me stress at once that the term 'radio wave' does not necessarily mean that the source of the radiation is artificial. The Crab is a radio source; so, for that matter, are the remnants of the other observed supernovæ, but I would be the last to suggest that any cosmical intelligence is responsible!

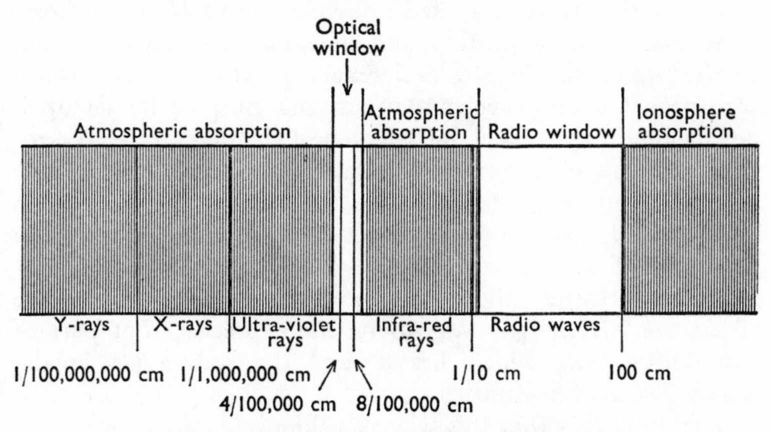

Fig. 67. The electromagnetic spectrum.

The whole range of wavelengths is known as the electromagnetic spectrum. I have shown it in Fig. 67, and it is clear that the visible range is very restricted—in fact it is even more restricted than the diagram indicates, because it is impossible to make a sketch to the correct scale. Before 1931 we had to draw all our knowledge from the limited range between 4,000 and 7,600 Å, and this meant that we were badly handicapped, rather as a pianist would be rather depressed at having to use a piano which has lost all its notes except those of the middle octave. (Again the scale is wrong; an octave includes about one-seventh the number of notes on a normal piano keyboard, but the range of visible light is much less than one-seventh, or even one seven-thousandth, of the whole electromagnetic spectrum.)

A further difficulty is caused by the Earth's atmosphere, which is irritatingly opaque to most wavelengths coming in from space. Visible light can pass through, of course—otherwise the sky would be featureless—but most of the short and longer wavelengths are stopped, as effectively as the beam of a torch will be stopped by a sheet of wood. Apart from the 'optical window', there is only one other major gap: the 'radio window'. Everything else is cut out.

The initial breakthrough was made by accident. In 1931 a young American of Czech descent, Karl Jansky, began some research on behalf of his employers, the Bell Telephone Laboratories. He was briefed to investigate problems of short-wave radio communication, and he was particularly concerned with 'static', the hissing and crackling which bedevils radio contact at short wavelengths. For this purpose he set up a strange-looking radio aerial, which he nicknamed the Merry-go-Round (parts of it were made from a dismantled Ford car). He picked up plenty of static, but there was something else too: a weak, steady hiss which appeared to come from the sky. The direction of the source could be estimated, and it moved from day to day. Eventually, Jansky found the answer. The hiss came from the Milky Way—or, to be more precise, that part of the Milky Way which lies toward the star-clouds in the constellation of Sagittarius.

It was known that this was also the direction of the centre of our star-system or Galaxy. Jansky, obviously, was interested. It seemed that he had picked up cosmical radio waves, and he published his work in 1933, probably expecting that it would cause general excitement. Nothing of the sort happened. His spectacular discovery caused about as much excitement as a piece of cake falling upon damp blotting-paper. Nobody seemed inclined to follow it up; and neither did Jansky. He produced a few more papers, and then turned his attention to other matters which had nothing to do with the sky.

There the matter rested for some years. An American amateur named Grote Reber built what was certainly the first intentional radio telescope, in 1937, and began to obtain results of immense significance; as well as confirming Jansky's discovery of radio waves from the Sagittarius star-clouds he also found sources in Cygnus, Cassiopeia and elsewhere—but his work, too, fell upon ears which were if not deaf, at least disinterested. There was even some scepticism about his work. The radio waves did not come from brilliant stars such as Sirius or Capella; they came from apparently blank parts of the sky, which was indeed a most peculiar state of affairs.

Then, during the war, British investigators led by J. S. Hey made another discovery which was to some extent fortuitous. It was found that some mysterious radiations were jamming some of our wartime radar equipment, and eventually the source of the trouble was tracked down not to the Nazis, but to the Sun. At this time the Sun had not long passed the peak of its eleven-year cycle of activity, and solar flares were common enough—whereas Jansky's earlier attempts to identify solar radio waves had been unsuccessful, because in the early 1930s, when he was experimenting with his pioneer aerial, the Sun had been 'quiet'.

Obviously Hey's reports could not be published at the time, because of military security, but when they were finally released they showed that radio astronomy had an almost limitless potential. Major 'radio telescopes' were designed, and were built. Much the most famous of them was the Jodrell Bank 250-foot dish, set up in Cheshire almost entirely because of the energy and dedication of Professor (now Sir Bernard) Lovell, who has really done for radio astronomy what George Ellery Hale, architect of the Palomar 200-inch optical reflector, did for visual astronomy.

The term 'radio telescope' is somewhat misleading, because an instrument of this sort is not really a telescope at all; it is in the nature of a wireless aerial, and one cannot look through it. It does not produce a visible picture of the object being studied, and all that one normally gets is a trace on a moving roll of paper. The only point of similarity between a radio telescope and an optical telescope is that both collect and focus radiations by using a curved reflector, though of course the radio telescope has a 'mirror' which is of metal rather than glass. Also, let me dispose of the idea that one is listening-in to noise from space. On many occasions 'solar noise' has been broadcast, sounding like the hiss of an enraged snake; but in fact the hiss is created inside the receiver, and is only one way of studying the long-wavelength energy coming in from beyond the Earth. Sound-waves cannot travel in vacuum, and for this reason alone there is no question that we could ever hear sounds from space. Moreover, not all radio telescopes are of the Jodrell Bank dish type. Many of them consist of long lines of aerials, and look nothing at all like telescopes of any kind. Each design has its own advantages, and its own disadvantages too.

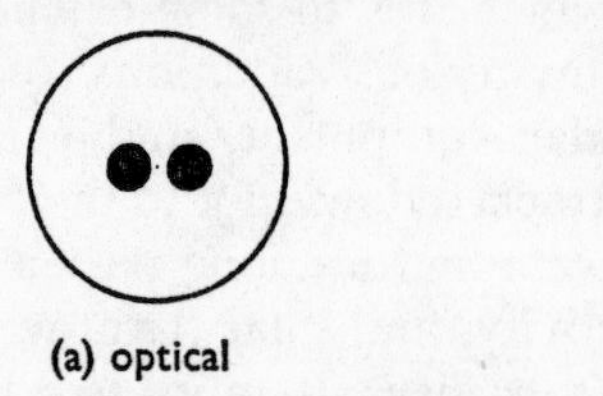

Fig. 68. Resolving power. Two optical sources will appear as clearly separated with an ordinary telescope, but (b) a radio telescope will blur two neighbouring sources, so that they cannot be divided.

One trouble about a radio telescope is that as compared with an optical instrument, it has very low resolving power. The diagram shows what is meant (Fig. 68). When you have two optical sources—two stars, if you like—they can be seen separately. If there are two radio sources, they tend to merge together, because they cannot be pinpointed so exactly, and there is bound to be an area of uncertainty. Neither can one tell whether the emissions are coming from one radio source or from several. Things are now much better than they used to be, but in the early days of radio astronomy research the areas of uncertainty were depressingly large.

The first success at identification came in 1949. Australian investigators managed to track down the positions of three radio sources accurately enough to link them with visual objects. Two of these were galaxies, well beyond our Milky

Way system. The third was none other than the Crab Nebula, which became known in radio parlance as Taurus A.

If the Crab were a radio source, what about other supernova remnants? The position of the strong source in Cassiopeia had already been tracked down by Professor Martin Ryle (now Astronomer Royal) and F. Smith at Cambridge; subsequently it was found that there were wispy patches of nebulosity in this position, spreading outward at 4,500 miles per second from an old explosion-centre, and it is now practically certain that a supernova blazed out there about the year 1700, though we could not see it because of intervening gas and dust—a sort of cosmical fog. Later came studies of the Cygnus Loop, another gas-cloud which seems to have been produced by a supernova which would have been on view around 65,000 B.C. if only there had been any astronomers on hand to observe it. Other cases are known, and it is now thought that a supernova remnant must necessarily be a source of radio waves.

Yet for a long time the mechanism of production of these waves remained a mystery. But for the Crab Nebula, it might well be a mystery still; but it is time to return to the events leading up to a supernova outburst. Not all stars become supernovæ; the Sun will never do so, for instance, because it is not massive enough.

Let us go back to the story of stellar evolution. As we have seen, a young star contracts and joins the Main Sequence (provided that it is massive enough for nuclear reactions to begin inside it), and for a long time it shines steadily, or virtually so. Just where it joins the Main Sequence, and how long it stays there, depends upon its initial mass. If it is comparable with the Sun it will build up a core of helium; when its available hydrogen is exhausted it will become a Red Giant, and different energy-producing processes will take over; then, when it has reached the end of its resources, it collapses into the White Dwarf condition, finally ending up as a dead globe. It will take a vastly long time to die, and there will be nothing catastrophic about its final extinction.

Now suppose that the original star is at least three and a half times more massive than the Sun. All kinds of complicated forces are operating; and if the helium core becomes too massive, the whole delicate balance of the star is upset. Heavier and heavier elements are formed, and as the core temperature continues to rise we meet with new complications; for instance we have to consider neutrinos, which are particles with no mass or electric charge, but which can carry off a great deal of energy.

The heaviest atomic nucleus which can be formed by continually increasing temperature and pressure is that of iron, which is produced when the internal temperature is above 3,000 million degrees. As the star evolves, the temperature at the core soars far above this limit, and at 5,000 million degrees a remarkable thing happens. Instead of building up into still heavier elements, the iron nuclei split up, and the process stops. The iron is converted back into helium; gravitational forces result in a sudden collapse of the star's core, and the outer layers, where the earlier nuclear reactions are still going on, are abruptly heated to something like 300 million degrees. (Remember that the surface temperature of our present-day Sun is a mere 6,000 degrees.) The result is a supernova outburst. Within a few seconds, the star emits as much energy as the Sun does in millions of years. Material is ejected—not only a small proportion of the star's mass, as with an ordinary nova explosion, but a high percentage of it. When the cataclysm is over, we are left with an expanding cloud; the remnant of the original star is composed mainly of neutrons, caused by the running-together of protons and electrons. Massive though it is, a neutron star may be much less than a hundred miles in diameter.

Much of this is still speculative, and different authorities have different ideas about what actually happens; also there are two definite types of supernovæ, which may not function in precisely the same way. The more luminous (Type I) have very little hydrogen in their outer layers, whereas Type II supernovæ have a good deal. Yet in any case we may be sure that these outbursts are on a scale grander than anything else in Nature, and at least we know more about them than we did a few years ago.

Of course, astronomers would welcome the chance to study a supernova outburst from relatively close range (not too close, however!) and we may have the chance at any moment; by the law of averages we are overdue for a fresh supernova.

Meantime, we have to satisfy ourselves with studying supernovæ in other galaxies, and fortunately this can be done. The classic example was that of S Andromedæ, which appeared in 1885 in the galaxy Messier 31, otherwise known as the Great Spiral in Andromeda—an independent system more than two million light-years away from us. The supernova became just bright enough to be visible with the naked eye, and was discovered around the same time by several observers. I particularly like the case of the Baroness de Podmaniczky, a Hungarian lady who was holding a house-party and had a small portable telescope set up on her lawn for the amusement of her guests. Just for fun, she was looking at the Andromeda Spiral when she announced that it looked rather different from usual. One of the guests, an astronomer who rejoiced in the name of de Kovesligethy, confirmed that there was something peculiar about the Spiral, and notified the authorities. In fact it may well be that the star had been first seen slightly earlier, but I am sure that S Andromedæ remains the only supernova to have been accidentally discovered by a Hungarian baroness.

At the time, nobody could be sure whether or not the star was connected with the Spiral; this was well before the revelation that spirals are separate galaxies rather than parts of our own system. However, many supernovæ have since been seen in outer galaxies, and their light-curves have been studied. The outbursts are powerful enough to be seen over vast stretches of space, and if we make the very reasonable assumption that a supernova in an outer galaxy is about as luminous as a supernova in our own Galaxy we can use these colossal explosions to make distance-estimates. A typical supernova in a remote galaxy is shown in Plate III.

But of all supernova remnants, the Crab is the most informative, and it contains the only neutron star which we can actually see. This brings us on to the problem of those extraordinary objects known as pulsars.

The story of pulsars really began in 1967, when radio astronomers at Cambridge were carrying out surveys with the help of new equipment. The research team was led by Professor Anthony Hewish, and included a graduate student, Miss Jocelyn Bell. It was Miss Bell who made the classic discovery. She found a weak radio source which appeared to be fluctuating very quickly indeed, almost as though it were 'ticking'. Originally she took it to be some kind of artificial interference, but further observations showed that it kept to the same position in the sky, in the constellation of Vulpecula (the Fox). What could it be?

The Cambridge team became very interested indeed, and the strange radio source repaid them by vanishing from the records. Then, one day in late November, Miss Bell announced: "It's back." At once arrangements were made to study it closely, and the results were quite staggering. The signals came as a regular succession of pulses at an interval of just over one second. They were absolutely regular, and utterly unlike any radio source found before.

It is fair to say that the researchers were taken aback. For a while it was even thought that they might have picked up artificial transmissions from some remote alien civilization, and the 'L.G.M.' or Little Green Men theory was not finally abandoned for some days. An official announcement was postponed until February 1968; by then it had become obvious that whatever it might be, the source was natural. A new term, 'pulsar', entered astronomical language, and the hunt was on. By the following autumn nine pulsars have been tracked down from Cambridge, Jodrell Bank and elsewhere, and at the time when I write these words (December 1973) the grand total is over sixty.

The quality of the radio pulses showed that the sources were definitely contained in our Galaxy rather than being external. It was also found that though the pulses were regular, they were in general slowing down slightly, at least for most of the pulsars; and the objects responsible for the transmissions had to be small by cosmical standards, with diameters measured in a few tens of miles rather than thousands or millions. The first sensible theory suggested that they might be White Dwarfs, vibrating rapidly, but it soon became apparent that no White Dwarf could vibrate quickly enough. Some pulsars send out their signals at a rate of more than four per second; and even though a White Dwarf is a small star, it is not small enough to behave like that. Indeed, there is one known pulsar with a period of only one-thirtieth of a second. The only type of object which can fit the facts is a neutron star.

Once again we come back to the Crab Nebula; where would astronomers be without it? It was examined anew, and a pulsar was found, vibrating in a period of a mere 0·03 seconds. Then, in 1969, astronomers in Arizona found that in the same position as the pulsar there was a faint star, which had previously been identified as the remnant of the 1054 supernova, but had not been seen to be variable. It proved to be flashing —and its period was 0·03 seconds, the same as that of the pulsar. Coincidence could be ruled out. For the first time a pulsar had been identified with an optical object; so far it remains the only case. Of course, the flashing star is extremely faint. It never rises above the fifteenth magnitude, so that large telescopes are needed to show it. Other workers, using rocket equipment to send their instruments above the inconveniently opaque atmosphere of the Earth, have established that the pulsar is also a source of X-rays.

It is tempting to believe that all neutron stars are supernova remnants, and this may well be true; the fact that we cannot see others visually is not significant, because the Crab pulsar (NP 0532, to give it its official catalogue number) is exceptionally close—a mere 6,000 light-years away, as we have noted—and only came into existence, *as* a pulsar, a thousand years ago as reckoned from our vantage point on the Earth. If the others are more remote and are also older, so that some of their initial energy has been used up, we could hardly expect to see them. It is also worth remembering that the Crab pulsar has the shortest period known, which again indicates its comparative youth. There are many other fascinating pulsars now under surveillance; Centaurus X-3, for instance, which is a binary system, and includes an X-ray pulsar.

A neutron star must be a weird object by any standards. It may have a diameter of between five and one hundred miles; there may be a thin solid crust in which sudden disturbances take place, producing 'starquakes' which show up by sudden alterations in the pulsars' periods. According to some calculations, the density at the centre of a neutron star is such that a cubic inch of it would weigh over four thousand million tons. Compared with this, even White Dwarf material is flimsy. So far as the pulsations are concerned, the current theory is that a neutron star has a very strong magnetic field, and is also rotating quickly. The magnetic pole is not coincident with the pole of rotation, and the result is that there is a kind of 'searchlight beam' effect; every time we pass through the 'beam' we receive a radio pulse. In fact, the pulses are produced not because the neutron star is vibrating, but because it is spinning round with amazing speed—thirty times every second in the case of the Crab pulsar.

Whether these ideas are valid or not we cannot yet be sure, but few astronomers now doubt that pulsars are neutron stars, representing the end products of supernova outbursts. This alone is a major step forward. Even though pulsars have been known for less than a decade, we have at least a reasonable understanding of them. Unquestionably they provided one of the main astronomical talking-points of the 1960s.

Chapter Thirteen

BLACK HOLES

IT MIGHT WELL be thought that neutron stars must represent the ultimate in celestial oddities. Yet this may not be so. During the past year or two much has been heard of what we call Black Holes—and if they really exist, they are stranger still. I have given them a separate chapter because I am acutely conscious that what I propose to say may be shown to be completely wrong within a matter of months. All I can do is to present the theories which are being so widely discussed at the present time.

According to the laws of physics, at least as we usually interpret them, a star with more than about ten times the mass of the Sun could never become a White Dwarf or an ordinary neutron star. During the abrupt shrinking, gravitation takes over, and the material is crushed together until the escape velocity becomes greater than the velocity of light (a concept which was, incidentally, discussed by the French astronomer Laplace as long ago as 1798). This means that the object is cut off from the rest of the universe, because not even light can get away from it. We have a very small, super-dense object—sometimes called a collapsar, from 'collapsed star'—surrounded by a kind of forbidden zone which makes up the black hole. The boundary of this forbidden zone is termed the event horizon. It has even been suggested that the original body may be crushed out of existence altogether; the black hole will still exert a gravitational pull on other bodies, so that it will be detectable, but obviously it cannot be seen.

It was in 1971 that the idea of black holes really came to the fore. This time it was not the Crab Nebula which provided the impetus, but an ordinary-looking star called Epsilon Aurigæ, close to the brilliant Capella.

I have already referred to the three 'Kids' which make up a naked-eye triangle. One, Eta Aurigæ, is quite ordinary; the second, Sadatoni or Zeta Aurigæ, is an eclipsing binary with a period of 972 days. Epsilon Aurigæ is even more remarkable.

The variations in light are quite noticeable, and were discovered by an astronomer named Fritsch as long ago as 1821. Normally, the star shines with a magnitude of 3 (slightly brighter than Megrez, the faintest of the seven stars in the Great Bear pattern), and the bright component, which sends us virtually all the light from the system, is definitely a yellow supergiant, with an F-type spectrum and a luminosity of about 60,000 times that of the Sun. The diameter has been given as 150 million miles, though I would hesitate to suggest that this figure is very accurate. The mass is 35 times that of the Sun, which by stellar standards is very high indeed. The secondary component cannot be seen directly, because its surface is so cool that it radiates only in the infra-red. Generally, the only spectral lines seen are those of the primary, though during an eclipse some extra dark lines appear.

An eclipse occurs every 27·1 years, or, more accurately, 9,898·5 days. The bright component is gradually covered by the invisible secondary, and the magnitude drops to just below 4. A complete eclipse lasts for over 700 days, and during the central 330 days the light remains steady at its minimum value. This, of course, provides a means of measuring the diameter of the secondary, and astronomers worked out that it must be of the order of 2,000 million miles—that is to say, big enough to swallow up the orbits of all the planets in our Solar System out to beyond Uranus. It would be the largest star known, with a calculated mass of 23 times that of the Sun. Presumably we were dealing with a very young star, still contracting toward the Main Sequence and not yet hot enough at its core for nuclear reactions to begin, whereas the primary had already passed through its Main Sequence stage and had passed into the giant branch to the upper right in the H-R Diagram.

This idea was challenged by two American astronomers, A. G. W. Cameron and R. Stothers, who suggested that instead of being young, the invisible secondary must be very old. This alters the whole picture. If the primary has already evolved off the Main Sequence (as seems definite) it is assumed that the secondary did so even earlier, and suffered an 'implosion', turning into a collapsar surrounded by a black hole. In this case, the infra-red radiation which we receive from Epsilon Aurigæ is due to a cloud of solid particles moving round the collapsar at a distance of around fifteen thousand million miles from the centre of the black hole. Over the course of time, many of these particles will spiral inward and enter the black hole, so that they too will disappear from our ken. Remember, nothing that is swallowed up by a black hole can ever emerge again.

It is a fascinating picture. By no means all astronomers accept it, but in any case Epsilon Aurigæ is an exceptional system, innocent though it looks.

Another 'suspect' is to be found in Cygnus. Here we have a supergiant star, HDE 226868, which has a mass about 30 times that of the Sun. It is certainly a binary, but the companion has never been seen, and is thought to have a diameter of only about sixty miles—and yet its mass is fifteen times as great as the Sun's. X-rays come from the system, and if these are due to the secondary they can be explained by supposing that the material falling in toward the black hole is accelerated to very high velocities, building up energy which is converted into heat.

Cygnus X-1, as it is called, has been carefully studied of late, and more and more authorities are becoming disposed to accept the black hole idea. With its high mass, the density of the collapsar might not be so great as that of a neutron star (in passing, if the Sun were to produce a black hole its diameter would have to be reduced to about 3 miles). Yet agreement is not complete. According to an alternative theory, there is no collapsar and no black hole in the Cygnus X-1 system, but merely two stars whose very powerful magnetic fields become entwined as the stars spin round; the magnetic lines of force are periodically broken and then reconnect, with the release of energy and the production of X-rays.

Nothing in modern astronomy seems to have caught hold of people's imagination as strongly as the black hole concept. With great diffidence, I suggest that there have been some rather wild speculations, and that there is a real need for caution. It has been suggested, for instance, that there is a huge black hole near the centre of our Galaxy, which will eventually swallow up every star and every planet; that space-travellers of the future, journeying between the stars, will have to be very careful to avoid falling into a black hole whose event horizon cannot, of course, be seen; and that it might be possible not only to utilize the power of black holes, but even to create them on a miniature scale, with potentially disastrous results. One particularly strange suggestion was that the so-called Siberian Meteorite of 1908 was nothing more nor less than a tiny black hole which hit the Earth, causing local devastation, and then passed right through the globe, emerging on the far side and continuing blithely on its way!

Then, too, there have been speculations about what would happen to an astronaut unwise enough to pass over an event horizon into a black hole. If he could survive the conditions there (such as super-violent tidal forces) it has been claimed that he might go through some sort of vague funnel and enter a completely different universe. Also, can a black hole reappear suddenly in another area as a 'white hole'?

Quite obviously, we have entered the realm of science fiction, and despite all these bizarre ideas we must not forget that up to now we have no certain proof that black holes exist at all. It is certainly premature to suppose that they are common, and contain an appreciable part of the mass of the entire universe. There may be a completely different explanation, and at the moment it is best to adopt a policy of 'Wait and see'.

Meantime, we may at least be confident that our Sun will not become either a pulsar or a black hole. It will end its brilliant career much less spectacularly, as a White Dwarf, and it will not change markedly for several thousands of millions of years yet. But it is intriguing to note how astronomy has developed; there is more than a casual connection between Tycho Brahe's supernova of 1572 and the pulsars and black holes of 1974.

Chapter Fourteen

CLUSTERS OF STARS

LOOK INTO THE eastern sky during any clear evening in early autumn, and you will see what appears to be a patch of shining haze. Look more closely, and you will see that the patch is really made up of stars, crowded so close together that they seem to be almost touching. If the sky is dark enough, anyone with normal eyesight should be able to make out seven individual stars in the group, and binoculars will reveal many more. The group is known officially as M.45 or the Pleiades cluster, though the popular name for it is the Seven Sisters.

The cluster is very easy to locate. It lies in Taurus, the Bull, some way from the bright red star Aldebaran. The diagram on the next page shows its shape, though in fact the Sisters are so noticeable that they can hardly be mistaken.

There has never been much doubt that the cluster is a real one, and is not due to accidental lining-up of stars at different distances from us. There is even a reference to it in Homer's *Odyssey*: Ulysses "sat at the helm and never slept, keeping his eyes upon the Pleiads". All the ancient astronomers, such as Ptolemy, described them, and there are various legends, about them. One version tells us that the Sisters were seven nymphs, who were strolling placidly through a forest when they were pursued by the hunter Orion. Orion was attracted by their beauty, and it is reasonable to assume that his intentions were anything but honourable; but just as he was about to overtake them, the King of Olympus changed them into doves and transferred them to the safety of the sky. (Orion's views about this episode are not on record.)

The brightest member of the cluster is Alcyone, which is of the third magnitude. Next in order of brilliance come Electra, Atlas, Merope, Maia and Taygete, while Pleione, Asterope and Celæno are on the fringe of naked-eye visibility; Pleione, the interesting 'shell star' described earlier, is however very close to Atlas, and the two may be hard to separate (Fig. 69). There have been suggestions that one of the Pleiades stars has faded

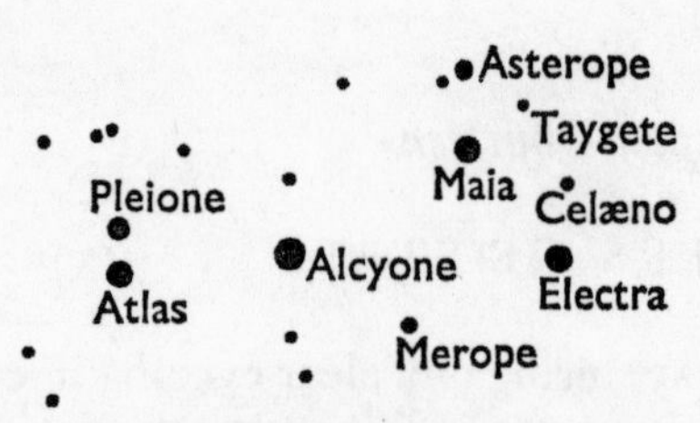

Fig. 69. The Pleiades: showing the main naked-eye stars, of which Alcyone (magnitude 3) is the leader. Try seeing how many you can count without optical aid. On a good night you should manage at least seven.

since ancient times, but I rather doubt this, though it is true that Pleione is somewhat variable. Some years ago I conducted a survey in *The Sky at Night* programme on BBC Television, and found that the average number of stars seen individually really is seven; but some people could see more. The record is usually said to be held by Eduard Heis, a last-century German astronomer, who totalled nineteen; but Terence Moseley, an amateur astronomer who lives in Armagh, has seen fifteen. Let me add, however, that this is exceptional. My own score has never exceeded a modest eight.

In the mid-19th century Johann Mädler, a German astronomer who will always be remembered as the first man to draw a really good map of the Moon, commented that the stars in the Pleiades did not show any motions relative to each other, and went on to propose that Alcyone might be the central star in the Galaxy, with all other stars (including the Sun) revolving round it. The idea became quite popular for a while, no doubt because of Mädler's excellent reputation, but it was quite unfounded. The Pleiades represent a particularly good example of an open or loose cluster—in this case a group of hot white stars embedded in very thin gas. There are over three hundred of them all told, contained in an area about fifty light-years in diameter. The distance from Earth is only just over 400 light-years, which explains why the cluster is so conspicuous. The best view of the Pleiades is obtained with binoculars. Use a telescope, and the field will be too small to show the whole cluster at once, so that much of the beauty will be lost.

There is little doubt that the stars in the Pleiades have a common origin, and were formed in the same way at about the same time—fairly recently on the cosmical time-scale, because the leaders are very luminous stars of spectral type B, and have not yet had time to evolve off the Main Sequence. A loose cluster of this kind is not a really stable system, and will eventually be disrupted by the gravitational pulls of other stars, so that it will lose its identity. The late Walter Baade, who was responsible for a complete revision of our estimates of the size of the universe, once wrote that the Pleiades cluster could not persist, *as* a cluster, for more than about a thousand million years in the future.

Open clusters are by no means uncommon, and several others are visible with the naked eye. In particular, note the Hyades, also in Taurus. They lie round the brilliant K-type star Aldebaran, and are just as easy to find as the Pleiades, though they are not so striking. They are decidedly overpowered by the strong orange light of Aldebaran, and in some ways this is a pity, since Aldebaran itself is not a true member of the cluster at all. Here we have a genuine line-of-sight effect; Aldebaran lies roughly half-way between the cluster and ourselves. When you look at the rather V-shaped form of the Hyades, it is worth remembering that Aldebaran is as far from the cluster as we are from Aldebaran.

The stars in the Hyades are relatively scattered, and the cluster is of different type from the Pleiades. The leaders are, in the main, Red Giants, and the very brilliant B-type stars characteristic of the Pleiades are absent, so that evidently the cluster is older.

The proper motions of the Hyades stars can be measured, and an interesting fact emerges. All of them seem to be converging toward a point in the sky about five degrees to the east of Betelgeux. Here we have a perspective effect; the stars are not really closing in on each other, but are moving through space in paths which are more or less parallel. Aldebaran, on the other hand, is moving almost at right angles, so that in the distant future it will no longer masquerade as a member of the cluster. Because the Hyades are apparently converging—and are, incidentally, receding from us—they will eventually give the impression of a smaller, more compact group, but these changes are so slow that they are quite inappreciable to the naked eye even over vast periods of time.

Another famous northern-hemisphere cluster is Præsepe in Cancer, popularly known as the Beehive or the Manger. Cancer itself is in the Zodiac, but it is decidedly obscure, and contains no stars much brighter than the fourth magnitude; it looks not unlike a very dim and ghostly Orion, but it is easy to find, because it lies between the Twins, Castor and Pollux, to one side and the Sickle of Leo on the other. (Here too we have a legend, according to which Cancer was a sea-crab which attacked the hero Hercules as he was battling with a particularly peevish monster. Hercules not unnaturally trod on it, but domestic quarrels among the Olympians led to its being transferred to the sky.) Præsepe lies near the middle of the constellation, flanked by two dim stars which are known as the

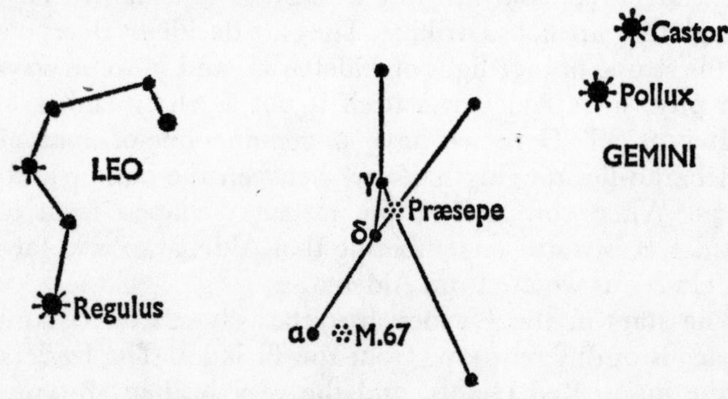

Fig. 70. Positions of the two clusters in Cancer. M.44 (Præsepe) is an easy naked-eye object, while the old cluster M.67 can be seen with binoculars. Cancer is a dim constellation, but is easy to find because of its position between Regulus on the one side, and Castor and Pollux on the other.

Aselli or Asses: Delta Cancri (magnitude 4) and Gamma (5). The cluster is an easy naked-eye object, though mist or moonlight will drown it. The leading stars are of mid- or late-spectral type, so that Præsepe is more closely related to the Hyades than to the Pleiades; it is fairly old, and seems to contain no nebulosity at all. The best way to view it is to use a very low magnification and a very wide-field eyepiece on your telescope.

Præsepe was the 44th object in Messier's catalogue. In the same constellation there is another famous open cluster, M.67, which is easy to see with binoculars near the inconspicuous star Acubens or Alpha Cancri (Fig. 70). It is a long way from the main plane of the Galaxy, and contains 500 members; it is also very remote—its distance from us is over 2,500 light-years, as against less than 600 light-years for Præsepe. The main point of interest, from the theoretical point of view, is that it is thought to be one of the very oldest of all open clusters, with an age of the order of 5,000 million years; it has avoided disruption only because of its remoteness from the galactic plane. The great French astronomer Camille Flammarion commented that it looked "like a sheaf of corn", which is by no means a bad description of it.

Even more beautiful telescopically are the twin clusters in Perseus, marking the sword-handle, easily found by using two of the stars in the W of Cassiopeia as pointers. The twins are in the same low-power telescopic field, and are true neighbours (Plate IV). The distance is thought to be 7,000 light-years, and the clusters may not be much more than a million years old. It is a great pity that they are so far away. Were they closer to us, they would provide a spectacle beside which the Pleiades would pale into insignificance.

Another striking cluster is to be found round Kappa Crucis, in the Southern Cross—unfortunately never visible from Europe. It has been nicknamed the Jewel Box, because it contains stars of various colours. A small telescope will show it, and it really does come up to expectations.

It would take a long time to describe even a few of the many open clusters which can be seen with modest instruments; I have given details elsewhere. Incidentally, it is ironical that when Charles Messier drew up his famous catalogue, he was not really concerned with clusters or nebulæ at all. He was a comet-hunter, and found that he was wasting time in checking upon faint, hazy patches which looked very like telescopic comets; so he catalogued them as 'objects to avoid'. It is ironical that while we still use the M-numbers, almost everyone has forgotten about his comet discoveries. For some reason or other Messier did not include the twin clusters in Perseus, and of course the southern objects such as the Jewel Box never rise above the horizon in France, where he carried out all his observational work. Much later, toward the end of the nineteenth century, J. L. E. Dreyer drew up a New General

Catalogue of clusters and nebulæ, containing many more objects than Messier saw, and the N.G.C. numbers are now used in preference. For instance, Præsepe is known as N.G.C. 2632 as well as M.44.

We also have stellar associations or 'moving clusters', which do not show up as compact groups. The best example is the Plough. Of the seven famous stars, five are moving through space in the same direction at the same rate, as we have already noted, so that they will keep the same relative positions for a very long time, and a few more stars belong to the same group. Generally speaking, a stellar association is defined as a number of stars of similar type and motion in the same region of the Galaxy; there is, for instance, an association made up of a very spread-out group of O and B stars in the area of Scorpio and Centaurus, and there is another round the highly luminous Zeta Persei. Here we have over fifteen hot stars, moving in a way which suggests expansion from a definite centre; the rate of expansion is about 8 miles per second, and it has been calculated that the expansion began about 1,300,000 years ago. It is reasonable to suppose that this is also the age of the stars in the association—which is hardly surprising, since all show O or B-type spectra, and must be young stars squandering their energy at a furious rate.

Associations of this kind are particularly important because they can tell us a great deal about the ages and past careers of the member stars. Astronomers all over the world are busily studying them; special mention should be made of the pioneer work carried out in the Soviet Union by V. A. Ambartsumian.

Let us now turn from the stellar associations, which are scattered so widely that they cannot be recognized on sight, to the globular clusters, which are as different as they can be. The two most spectacular examples lie in the far south of the sky, and can never be seen from Europe. One is Omega Centauri, not so very far from the Southern Cross; the other is 47 Tucanæ, in the rather dim constellation of the Toucan. Each is visible to the naked eye, and in a telescope each is showed to be a superb 'ball' of stars, strongly condensed toward the centre.

In the northern hemisphere, the brightest globular cluster is M.13 Herculis. It is just visible without optical aid on a clear night, but is none too easy to find unless its position is known accurately, particularly as Hercules itself is rather ill-defined. The cluster lies between Zeta and Eta Herculis, rather closer to Eta. If you cannot see it with the naked eye, the best course is to use a low power on your telescope and sweep from Zeta toward Eta; you should then find the cluster without much trouble, and, incidentally, it is obvious in binoculars. With a moderate telescope it is a fine sight, and many individual stars can be seen in its outer portions. It has been known for a long time; Edmond Halley, of comet fame, discovered it as long ago as 1714.

Globular clusters are comparatively rare. Only just over a hundred are known in our Galaxy, and the full total may not be more than about twice as great. Moreover, most of them are faint, and even Omega Centauri is not really striking with the naked eye. This is not because they are really dim, but because they are very remote. Omega Centauri, which is admittedly rather above the average, has a total luminosity of perhaps a million Suns, and lies at a distance of 22,000 light-years; M.13 Herculis is somewhat further away. Plate V shows another splendid globular, M.5 in Serpens, at 27,000 light-years from us.

The distribution of the globulars is rather unexpected. They are not spread uniformly around the sky, and most of them are in the south; there is a well-marked concentration toward Scorpio and Sagittarius. As we will see later, this apparently lop-sided distribution is due to the fact that the Sun lies well away from the centre of the Galaxy instead of being near the middle, as Herschel had believed. Their distances have been measured by making use of some of the convenient variable stars which provide us with information simply by the way in which they fluctuate. Most globulars are reasonably rich in RR Lyræ stars, though, suprisingly, M.13 Herculis contains very few.

The most powerful stars in globular clusters are red super-giants, and it follows that the systems are relatively old. They make up a sort of 'outer framework' to the Galaxy, and Omega Centauri is actually the closest of them.

The central crowding is real, and near the middle of a globular the average distance between individual stars is much less than in our own part of the Galaxy. The actual distances are still great, and collisions can hardly ever occur, but any one star is almost bound to have several more within a couple of light-years of it. Moreover, many of these will be very luminous. Suppose that we could take a trip to a planet moving round one of the stars near the middle of Omega Centauri or M.13? What sort of a sky would we see, and would it be very unfamiliar to our eyes?

The answer is a resounding 'yes'. On our imaginary planet, the scene will be magnificent. Instead of a few brilliant stars, there will be many thousands; probably at least thirty will shine more brilliantly than Venus does to us, and it is quite on the cards that several will rival our Moon, so that instead of appearing as twinkling points they will show real disks. There is also a strong chance that they will be red, since—as we have seen—the red supergiants predominate. There can be no true 'night' on such a world. The glare of the stars will give much more light than our full moon, and darkness will be unknown.

If we go a step further, and suppose that the planet is inhabited by a race of beings who share our own interest in the universe, we can see that certain difficulties will have to be faced. Admittedly, the close stars will be well displayed, but fainter objects will be drowned in the general glare, so that it will be a hopeless task to try to study anything outside the cluster.

The proper motions of the stars will be more marked than with us, since the stars in a globular are moving in paths round the centre of gravity of the system, and even over comparatively short periods these shifts will be noticeable. No constellation will keep its outlines permanently, and star-maps will have to be re-drawn completely every few centuries. Moreover, the apparent magnitudes of the stars will change. A red supergiant which passes within, say, a light-year of our imaginary planet will dominate the night sky; as it passes its point of closest approach and begins to recede once more, its disk will shrink and its glare will dwindle into a soft glow. There is also the fact that supergiants tend to be unstable and to vary in luminosity, so that magnitude changes would be caused in this way as well. No narrow-minded astronomer living in a globular cluster would be able to maintain that the skies are unchanging, as our own philosophers believed only a few centuries ago.

It is interesting to speculate as to the theories which might be held by cluster-dwellers. Probably they would believe their own system to be the only one, and they might well regard the universe as being very limited in size. They might have shrewd suspicions that other systems could lie beyond their local globular, but proof would be very hard to obtain.

Of course, we have no idea whether there are any inhabited planets inside globular clusters, but there seems no obvious reason why not. Each globular contains many tens of thousands of stars, and no doubt some of these are very like our Sun, though from our own vantage-point we can see only the highly luminous leaders.

We can see, then, that there are associations and groups of all kinds, from the spread-out moving clusters to the open clusters such as the Pleiades, the more complicated systems of the Perseus sword-handle type, and the wonderful starry spheres which we call globular clusters. There is endless variety in the stellar universe.

Chapter Fifteen

NEBULÆ

WHENEVER THE SKY is dark and clear, and Orion is above the horizon, the unmistakable pattern of its bright stars catches the eye at once. Quite apart from the eminence of Betelgeux and Rigel, we have the prominent Belt; and close by the Belt there is the Hunter's Sword, which looks rather like a faint patch of luminous fog. Moonlight drowns it, but under good conditions it is easily visible without a telescope, and any pair of binoculars will show it well. It is the most famous of those objects which we call galactic or gaseous nebulæ, and it has been known ever since 1610, when it was described by an otherwise totally obscure astronomer named Nicholas Peiresc.

The word 'nebula' is Latin for 'cloud', and the term is by no means inappropriate. (Certainly it is better than many others we find in astronomy; for instance there are no seas on the Moon, and no canals on Mars.) Even a modest telescope will show details in the Sword of Orion. In its midst there is the famous multiple star Theta Orionis, known as the Trapezium, and there are other stars too, giving the impression that they are lighting up the gas and making it shine. There is much truth in this, though we know now that it is only part of the story.

As we have noted, Charles Messier catalogued over a hundred objects in his famous list published in 1781. Not all of them were gaseous nebulæ. Many were star-clusters, and there were also some patches which seemed to be resolvable into stars. Of these, the most famous was the 31st object in his catalogue—the Andromeda Nebula, which had been recorded in the tenth century by the Persian astronomer Al-Sufi and which is dimly visible with the naked eye (though, for some unexplained reason, it was totally overlooked by the usually eagle-eyed Tycho Brahe). Increased telescopic power showed that there were two definite classes of nebulæ, the irresolvable —such as the Sword of Orion—and the resolvable or starry.

Using instruments which were much the powerful of their time, William Herschel set out to catalogue the various nebulous objects, and in 1786 he published a catalogue of more than a thousand of them. He mentioned "nebulosity of a milky kind" near Theta Orionis, and in 1789 he described it as "an unformed fiery mist; the chaotic material of future suns", which was remarkably near the mark. Later, he said that this self-luminous material "seemed more fit to produce a star by its condensation, than to depend on the star for its existence". Evidently he had more than a suspicion that the gaseous nebulæ might be stellar birthplaces; and he also speculated as to whether the resolvable nebulæ might be separate systems, far beyond our own.

We now know that this idea was correct, even though it fell into disfavour for many decades after Herschel's death in 1822. M.31 in Andromeda is not a nebula; it is a galaxy, and the same applies to the other resolvable patches, though not all of them are spiral in the same way as M.31. Nowadays the term 'spiral nebula' has been dropped. When referring to a nebula, we mean a cloud of dust and gas.

The first definite proof that irresolvable nebulæ are gaseous came on 29 August, 1864, with some brilliant work by the pioneer spectroscopist Sir William Huggins. When Huggins looked at the spectrum of a small nebula in Draco, he found that instead of showing a rainbow band crossed by dark lines there were bright emission lines only—and not many even of these. This was proof that the nebula must be composed of shining gas. (To be accurate, Huggins' object was a planetary nebula, which is really a star surrounded by a huge gaseous shell and is quite different from the Sword of Orion; but the same principles apply.)

Let us look more carefully at the Orion Nebula, M.42 (or, if you like, N.G.C.1976). Its distance from us is about 1,500 light-years, and it is extremely large. The diameter is of the order of 25 to 30 light-years (Fig. 70). This means that it could cover the distance between ourselves and the bright blue star Vega. Yet the density is incredibly low, and corresponds to what we normally call a laboratory vacuum. I am reminded of a comment made to me quite recently by Dr. David Allen, of the Royal Greenwich Observatory. If you could bore a one-inch core right through the nebula, the material you would collect would weigh no more than one new penny!

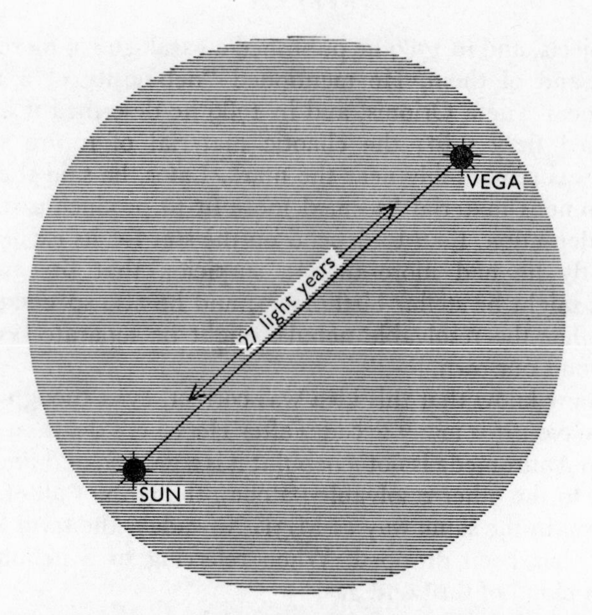

Fig. 71. Size of the main dark nebula in Orion. Its diameter is about 30 light-years; so that if the Sun lay near one edge, the other edge would extend beyond Vega.

There is plenty of dust in the Orion Nebula, and this makes stars inside it seem redder than they would otherwise be—because the shorter-wavelength blue light bounces off the dust-grains and is scattered. Some years ago, E. Becklin found an infra-red source in the nebula; it could not be seen visually, and was assumed to be a very young star which was still contracting toward the Main Sequence. More recently, work by Allen and his colleagues seems to show that Becklin's Star is in fact immensely luminous, but is hidden by the dust between it and ourselves. Other nebulæ may well contain stars of the same kind.

M.42 has a bright-line spectrum, as was first shown by the tireless Huggins in 1888, and this means that it is self-luminous to some extent. The stars inside it are extremely hot, with spectra of type O and B, and it is the ultra-violet radiation sent out by them which excites the nebular material to luminosity. Other nebulæ depend solely upon reflection; there is, for instance, the nebulosity in the Pleiades cluster, which is not at all easy to see visually though it is spectacular when photographed. The stars inside it are not so hot as those in the Sword of Orion, and so they cannot make the gas shine on its own account.

It is true to say that it is the gas which shines, and the dust which absorbs the light of stars inside the nebula. Hydrogen is much the most plentiful substance in the universe, and so it is not surprising to find that galactic nebulæ consist largely of it. There are other elements as well—oxygen, neon, nitrogen and argon, for instance—but hydrogen is dominant. Incidentally, there are also some spectral lines which caused a great deal of excitement when first discovered, because they could not be identified; it was thought that they must be due to a completely new element, which was named nebulium. There was an interesting precedent, since helium had first been found in the spectrum of the Sun, and had not been tracked down on Earth until a quarter of a century later. Nebulium remained a mystery until 1927, but then I. S. Bowen, in America, showed that it was due to nothing more fundamental than common elements (mainly oxygen and nitrogen) in unfamiliar states. This was something of an anti-climax, and nebulium, as an element, was rather sadly deleted from the records.

There can be no doubt at all that fresh stars are being formed inside galactic nebulæ, and much of our information has been drawn from M.42—not because it is unusual, but because it is so bright and easy to study. I have already mentioned FU Orionis, the very young star which was caught in the act of blowing away its dust-cloud and brightening up by several magnitudes. Moreover, the nebula contains over two hundred irregular variables which are still in the pre-Main Sequence stage. The first of them, AF Orionis, was found as long ago as 1848, but nowadays the stars are known as T Tauri variables after the most famous member of the class.

There is also some fascinating work carried out some time ago by the American astronomers Blaauw and Morgan. Their attention was drawn to a faint O-type star, AE Aurigæ, which seems to have an exceptionally high velocity of something like 80 miles per second. It seems dim only because it is so far off, and on the stellar scale it is both energetic and young. If we trace its path backwards, we find that about two and a half million years ago it must have been in the region of the Orion Nebula. In almost exactly the opposite direction there is another O-type star, Mu Columbæ, which has a similar velocity and is about the same distance from the nebula, but is moving the other way. It has been suggested that some colossal disturbance took place two and a half million years ago, hurling AE Aurigæ and Mu Columbæ violently outward. To this may be added a third star, 53 Arietis; the agreement is not so good, but lies within the limits of uncertainty. Can it be that these three stars were once near neighbours—perhaps in that stage in their careers when they were still, broadly speaking, being formed—and that some outburst sent them on their different ways, so that by now they are so far apart that as seen from Earth one lies in Columba, one in Auriga and one in Aries? It is at least possible.

Now let us turn to T Tauri, and what I call "The Case of the Vanishing Nebula". The nebula itself lies near the Hyades, and was first noticed in 1852 by John Russell Hind, an observer who was particularly interested in minor planets and was searching for them with the help of a 7-inch refractor. He described it as "a very small, nebulous-looking object" close to a tenth-magnitude star. T Tauri, the star concerned, was a variable, but Hind was in no position to realize that there was anything peculiar about it.

The nebula was duly listed, and seemed to be perfectly normal, but in 1861 Heinrich d'Arrest, in Germany, found to his amazement that it had almost disappeared. Very large telescopes still showed traces of it, but by 1868 it had completely gone. Hind's nebula was officially classed as 'missing', but other developments were taking place nearby. In 1868 Otto Struve found nebulosity round a 14th-magnitude star close to T Tauri, and d'Arrest, who had previously looked carefully at the area, was certain that the nebulosity was new. It was recorded at various times until 1877, but then it too disappeared as completely as the hunter of the Snark.

Then, in 1890, Barnard and Burnham, using the 36-inch refractor at the Lick Observatory, rediscovered Hind's old nebula. It was visible, but a mere ghost of its former self, and was a difficult object even with this powerful telescope. Late in

1895 it vanished again, to reappear once more later on. Nowadays it is easily visible with large instruments, but its form is not the same as when Hind first reported it. Of course, it is not in Messier's list; its official catalogue number is N.G.C.1555.

Moreover, T Tauri itself is embedded in nebulosity—and this nebulosity too is variable. And as an extra puzzle, it was found that Hind's nebula yields emission lines, though T Tauri is a dwarf star not nearly hot enough to excite nebular gas to luminosity. The nebula is much more variable than the star, and it was suggested that changes in illumination could be largely responsible, so that whereas in Hind's day the gas was lit up by T Tauri some other material moved in around 1860 and cast what could be regarded as a shadow.

T Tauri is an infra-red source. Careful studies of it have been carried out in the infra-red by Martin Cohen during the past few years, and when he correlated his observations with my own visual observations of the star's fluctuations he found that there was a definite connection. There can no longer be any doubt that T Tauri and its kind are veritable infants on the time-scale of the universe.

Hind's object is not the only variable nebula known. Others are, for instance, associated with R Monocerotis, in the Orion area, and R Coronæ Australis in the southern hemisphere of the sky. Unfortunately all these nebulæ are so faint that they can be studied only by means of photographs taken with large telescopes.

There are plenty of gaseous nebulæ within the range of a moderate telescope. Look for instance at M.8, the Lagoon Nebula in Sagittarius, which is very remote (nearly 5,000 light-years from us) and is of the emission type; it is also a radio source, and contains a large number of T Tauri variables. Also in Sagittarius there is M.17, the Omega or Horse-shoe Nebula, which contains both bright filaments and dark obscuring material. Southern observers have the Keyhole Nebula, which surrounds the extraordinary variable Eta Carinæ and is much larger and more massive than the Sword of Orion, though it is also much further away from us.

I have said that nebulæ associated with very hot early-type stars become self-luminous (H.II regions, to use the technical term) while if the nearby stars are cooler the nebula is visible by reflection only. It may be added that some bright nebulæ, notably M.42, represent only parts of much larger areas of nebulosity. But suppose that there are no convenient stars to provide illumination? Presumably the nebula will remain dark —and this is exactly what we find.

A dark nebula betrays itself because it cuts out the light of stars beyond, just as a cloud of smoke will hide a distant street-lamp, and the outlines of the nebula will be traceable. For once Herschel made a mistake here; it is said that on coming across one of these patches during a sky-sweep, he explained "This surely, is a hole in the heavens!" Yet there is no chance of stars being absent in one particular direction, and there is no doubt that shielding nebulosity is the answer.

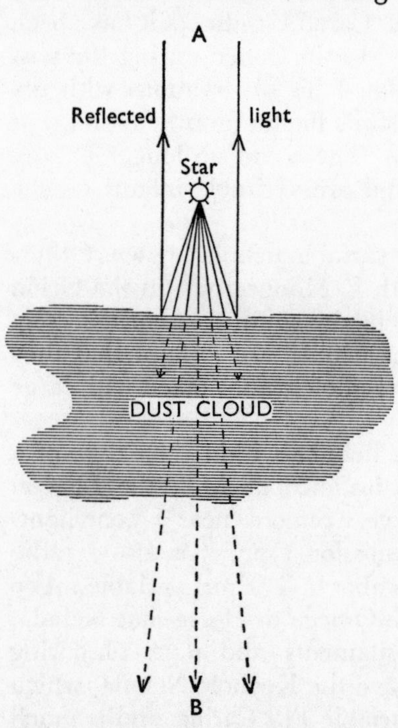

Fig. 72. From A, the dust cloud would be illuminated by the star, and would appear bright; from B, with no suitable star, an observer would see a dark obscuring mass blotting out all the objects beyond.

The best-known of these dark nebulæ, the Coal Sack in the Southern Cross, is never visible from Europe. It blots out an area of sky 8 degrees long by 5 degrees wide, and binoculars show it excellently. There is nothing so striking in the northern sky, although smaller obscuring clouds are to be found in Cygnus and elsewhere. Near the star Rho Ophiuchi, bright and dark nebulæ are seen side by side, giving a wonderful effect and there are similar nebulæ near Gamma Cygni (Plate VI).

There is no basic difference between a bright nebula and a dark one—and, moreover, much depends upon the angle of observation. In the diagram (Fig. 72), we have a nebula together with a powerful star. Seen from point A, the nebula would be bright; from point B it would be dark. By various indirect methods, distance-estimates can be made. The Coal Sack proves to be around 400 light-years away, and to be about 40 light-years across. If our Sun lay near the centre of a cloud of this size, familiar stars such as Alpha Centauri, Procyon, Sirius and Altair would also be inside it.

Different again are the so-called planetary nebulæ, which, as I have said, are neither planets nor nebulæ in the true sense of the term. The most famous of them is M.57, the Ring Nebula in Lyra, which was discovered in 1779 by an astronomer named Darquier who was using a telescope of only about 3 inches aperture. It is easy to locate, since it lies directly between Beta and Gamma Lyræ, close to the brilliant Vega (Fig. 73). Beta is the celebrated eclipsing binary, while Gamma is a normal star of the third magnitude. Look between them, using a modest telescope, and you will make out a dim patch. Increased power will show it as a luminous ring, not unlike a tiny shining bicycle-tyre. The name 'planetary nebula' is due to Herschel, who once thought it possible that the objects might be planetary systems circling other stars—though it was not long before he realized that this could not be the case.

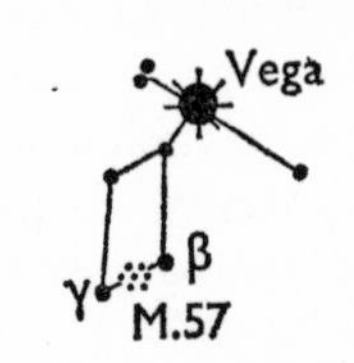

Fig. 73. Position of the Ring Nebula M. 57, the most famous of the planetaries. It lies between Beta Lyræ (the famous eclipsing binary) and Gamma Lyræ, so that it is easy to locate, though in a small telescope it is faint.

The Ring Nebula is made up of a central star of below the twelfth magnitude, surrounded by a spherical gaseous shell. When we look at it, we see more glowing matter at the edge than at the centre, so that we have the impression of a luminous ring. The overall diameter is rather less than one light-year, but other planetaries are larger. N.G.C.7293 in Aquarius is roughly twice the size of the Ring Nebula; if the Earth lay on one side of a planetary of this diameter, the opposite edge would extend well out toward Alpha Centauri.

Yet the planetaries are not nearly so massive as might be thought—because they are so rarefied. If you could take a cupful of air and spread it around a giant vacuum-flask five miles in diameter, the resulting density would be roughly the same as that of the gas in an average planetary nebula.

About five hundred planetaries are known, but most of them are very faint and far-away; even the Ring lies at over 1,400 light-years from us. Not all are symmetrical. The Dumbbell Nebula in Vulpecula, M.27, is one of the finest examples. It was first seen in 1764 by the indefatigable Messier, and has a central 12th-magnitude star with a featureless rainbow spectrum; it really does look like a dumb-bell when seen with a telescope of sufficient power. Then we have the Owl Nebula, M.97, in the Great Bear, not far from Merak, the fainter of the two Pointers to the Pole Star. This time we have two bright, smooth condensations symmetrically placed to either side of the central star. The Owl is large for a planetary, but is not an easy object—which is hardly surprising, since its distance is of the order of 10,000 light-years. Also worth looking at is H.IV.1—not catalogued by Messier—which lies in the same low-power field as the orange star Nu Aquarii. In very large telescopes it shows strange extensions which remind one a little of Saturn's ring system, though, needless to say, there is no real analogy.

All planetaries are expanding. Their average age is thought to be about 20,000 years; as the material spreads out it becomes fainter, and it has been estimated that after 100,000 years or so the shell will become too faint to be seen at all. The central stars have high surface temperatures of 50,000 degrees or so, and are well advanced in their life-stories, since they have completed their main nuclear burning and are on the way to becoming White Dwarfs. For instance, the explosion which produced the Dumb-bell Nebula is thought to have occurred between 3,000 and 4,000 years ago. Whether all stars of normal type become planetary nebulæ for a time is a matter for debate.

We have learned a great deal since Messier drew up his list of 'objects to avoid' during his searches for comets. His catalogue was a grand medley of gaseous nebulæ, galaxies, clusters and planetaries, to say nothing of the very first entry of all, the stellar wreck M.1—the Crab, which has told us so much. But of all these, it is perhaps the ordinary gas and dust clouds which hold the greatest fascination for us, for it is here that new stars are being created. Thousands of millions of years ago, our Sun was born in the same way.

Chapter Sixteen

THE GALAXY

THERE CAN BE few people who have not admired the beauty of the Milky Way, which stretches right across the sky from horizon to horizon. Ptolemy, in his *Almagest,* wrote that "The Milky Way is not a circle, but a zone, which is almost everywhere as white as milk, and this has given it the name it bears. Now, this zone is neither equal nor regular everywhere, but varies as much in width as in shade of colour, as well as in the number of stars in its parts, and by the diversity of positions; and also because that in some places it is divided into two branches, as is easy to see if we examine it with a little attention."

Ptolemy's account dates back nearly two thousand years, but as a description of the Milky Way as seen with the naked eye it could hardly be bettered. Starting in Cassiopeia, we can follow it through Perseus, Auriga and Gemini, past Procyon and Sirius down to the southern horizon as seen from England; it then crosses the Ship, the Southern Cross and Centaurus, and thence into Scorpio and Sagittarius, Aquila, Cygnus and back to Cassiopeia. The section between Carina and Sagittarius never rises above the European horizon, and this is unfortunate, since it is particularly rich; in Crux, too, we have the celebrated Coal Sack. However, the most brilliant part of the whole Milky Way is in Sagittarius. This at least is within range of European observers, though it is always rather low down; it is best seen over the southern horizon during summer evenings.

The Milky Way must have been known from very early times—after all, it can hardly be overlooked!—and various theories were put forward to explain it, some more logical than others. The Greek philosopher Parmenides of Elea, who lived toward the end of the sixth century B.C. held that "it is the mixture of the dense and the rarefied which produces the colour of the Milky Way". (To do him justice, he also claimed that the stars must be made of what he called "compressed fire", so at least he regarded them as self-luminous.) Anaxagoras of Clazomenæ, born about 500 B.C., believed that "the Milky

Way is the light of certain stars. For, when the Sun is passing below the Earth, some of the stars are not within its vision. Such stars, then, as are embraced in its view are not seen to give light, for they are overpowered by the rays of the Sun; such of the stars, however, as are hidden by the Earth, so that they are not seen by the Sun, form, by their own proper light, the Milky Way." This may sound rather involved, but I can assure you that it makes sense when you have read it through several times. Anaxagoras, of course, believed the Earth to be flat, and to him the Sun was a large red-hot stone. Yet he did think the Milky Way to be made up of stars, rather than a shining fluid.

When Galileo first began to use the telescope for looking at the sky, in the winter of 1609–10, he found that the Milky Way "is nothing else than a mass of innumerable stars planted together in clusters. Upon whichever part of it you direct the telescope, straightway a vast crowd of stars presents itself to view; many of them are tolerably large and extremely bright, but the number of smaller ones is quite beyond determination."

Here again we have an account which might have been written by any modern astronomer. The greater the light-gathering power of your telescope, the more stars you will see; but even a small instrument will show so many of them that to count them would be absolutely impossible. And at first sight, it might well be thought that the stars in the Milky Way are so crowded together that they can hardly avoid bumping into each other.

As so often happens in astronomy, appearances are deceptive. There is no real crowding; all we are seeing is a line of sight effect. Once again we come back to the work of Sir William Herschel, who was the first man to put forward a really scientifically-based explanation of the arrangement of the stars in our system.

Herschel knew that he had no hope of counting all the stars, so he decided to confine his main work to certain selected regions of the sky. This was the famous method of star-gauging, and it worked well. His final conclusions were not entirely accurate, but at least he was on the right track, and his work was based upon logical research rather than guesswork or (as in some previous cases) wild theology.

Herschel believed—wrongly—that the apparent brightness of a star must be a reliable guide to its distance from us, so that brilliant stars such as Sirius and Canopus would be closer than fainter ones such as Polaris. This sounds plausible, but we now know that the stars have so wide a range in luminosity that the principle does not hold good; as we have noted, Canopus, for instance, is a real searchlight seen from an immense distance. Herschel, of course, had no means of checking, because in his day star-distances were still unmeasured. If he had been right, it would follow that the regions of the sky which contained the most stars represented the greatest extensions of the stellar system—and this was where the star-gauging method came in. Obviously, there are more stars in and near the Milky Way than in other parts of the sky, but Herschel soon found that the percentage increase was greater for faint stars.

For instance, suppose that we take two telescopes—say a 3-inch refractor and a 10-inch refractor—and use them to study a selected area near the Milky Way, as well as another region in a barren part of the sky. The small telescope will show only fairly bright stars, and it will be found that the rich region will contain about four times as many stars, area for area, as the barren one. With the large instrument, which will show fainter stars as well, the ratio is bound to be much higher: probably about ten to one. In other words, faint stars are more numerous near the Milky Way than might be expected.

Herschel decided that the stellar system or Galaxy must be shaped rather like what he called "a cloven grindstone" or double-convex lens. This would explain the Milky Way effect easily enough. The diagram (Fig. 74) will show what is meant, though, to be honest, it is not drawn quite to Herschel's pattern. If the Sun (S) lies near the middle of the system, many stars will be seen if we look toward A or B, and this will produce the luminous band; fewer stars will be seen in the directions of C and D.

This was as far as Herschel could go, and the next real step was taken in 1904 by a Dutch astronomer, Jacobus Kapteyn, who showed that the proper motions of the stars are not entirely random; there is a tendency for a general drift in two special directions. Since all proper motions are tiny, the effects of this star-streaming are very slight, but Kapteyn

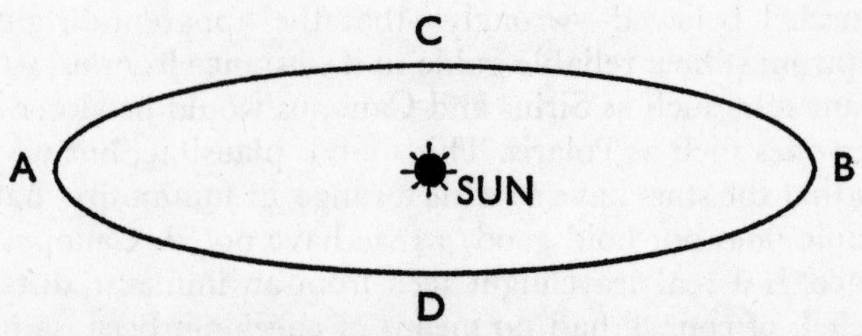

Fig. 74. The idea of the Galaxy as a system shaped like a double-convex lens, with the Sun near the middle. This is not exactly as Herschel pictured it, but it gives the general idea.

knew that the phenomenon must be real. Accordingly he decided to attack the problem of star distribution rather along the lines laid down by Herschel, but unlike his great predecessor he did not have to do all the work himself. Observatories all over the world co-operated in examining over two hundred regions for star counts, and from the results Kapteyn was able to draw up a new picture of the Galaxy. Again the Sun, with the Earth and other members of the Solar System, was thought to be rather close to the centre of the system.

The idea of a centrally-placed Sun was not unreasonable. A fly sitting on the hub of a bicycle-wheel will be able to see the rim both above, below, and to the right and left; similarly the Milky Way makes up a complete band in the sky. True, it is richer in some parts than in others, but this could well be coincidence. The key to the whole problem was eventually provided by the globular clusters; the man responsible was the great American astronomer Harlow Shapley, whose death only a few years ago was so sad a blow to his many friends (including myself).

We have already met the globular clusters, which are truly magnificent objects. Unfortunately there are not many which are really well seen from Europe, and the two brightest, Omega Centauri and 47 Tucanæ, are so far south that they never rise over either Europe or the main part of the United States. Shapley realized that the grouping was too marked to be sheer chance. Over a hundred globulars are known, and most of them lie to the south, particularly toward Scorpio and Sagittarius. This is not to say that the northern hemisphere has no globulars at all; remember M.13 in Hercules, for instance, which is a bright system. We also have some others, such as M.92, also in Hercules, and M.3 in Canes Venatici, the Hunting Dogs (not far from the Great Bear). All the same, Shapley was impressed by this lop-sided distribution. There had to be some explanation for it.

Shapley realized that the globulars found a kind of outer surround to the main Galaxy, and that they lie on the fringes of the system, which is why all of them are a long way away from us. At that stage their distances had not been measured, but luckily the RR Lyræ variables came to the rescue.

To recapitulate for a moment: the RR Lyræ stars are short-period variables, with periods of less than a day. Their fluctuations are regular, and all of them seem to have about the same luminosity, somewhat less than a hundred times that of the Sun. As soon as we can measure their apparent magnitudes, then, we can find their distances—allowing for various complications such as inter-stellar absorption, of course. This is what Shapley did with the RR Lyræ variables in the globular clusters, and for the first time he could work out not only their distances but also their sizes. The mystery was solved at once. The Sun lies well away from the centre of the Galaxy, and so we have a lop-sided view. This is where Herschel had been wrong.

Now we can draw up a better picture. There is a central nucleus, lying in the direction of the Sagittarius star-clouds, and the Galaxy has the flattened shape shown in Figure 75; it measures about 100,000 light-years from side to side, with the

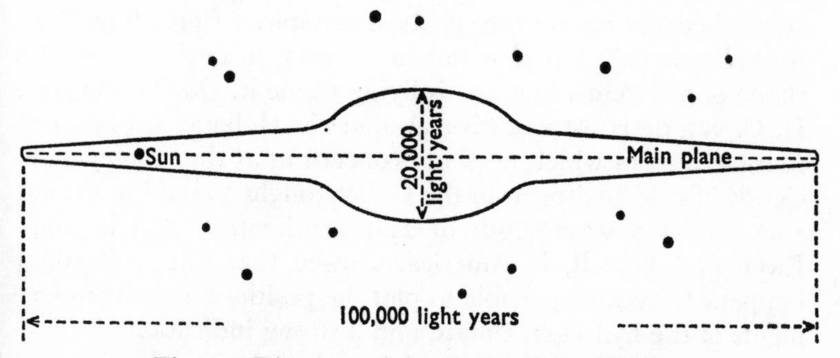

Fig. 75. Diagram of the shape of the Galaxy.

Sun lying about 32,000 light-years from the centre. The thickness of the nucleus is around 20,000 light-years. Surrounding the main system is the galactic halo, including the globular clusters and many isolated stars.

Nowadays we know the position of the centre of the Galaxy, but we cannot see it, because there is too much 'dust' in the way. And this brings us on to the general question of interstellar material, which has become so remarkably interesting during the past year or two.

The story really began in 1904, with some spectroscopic work by an astronomer named Hartmann. He was looking at the spectrum of Delta Orionis, the faintest of the three stars in the Hunter's Belt, which is a spectroscopic binary, so that its absorption lines show the characteristic to-and-fro displacement due to Doppler effects. Hartmann found that one line, produced by calcium, did not share in this movement, but remained obstinately in the same position. Clearly, then, it could not be a line due to the binary system itself, and could only be due to material lying between the star and ourselves. Since then many other cases have been found, and there is also direct proof of the presence of interstellar matter; I have already discussed the strong reddening of some remote stars. Not unexpectedly, the absorption of light is most marked near the main plane of the Galaxy.

It was also the presence of interstellar material—hydrogen gas, in this case—which led to proof positive that if our Galaxy could be seen 'from above' (or below) it would take the form of a spiral system. This had long been suspected, partly because of the way in which the bright stars are arranged in the sky and partly because many other galaxies are spirals (including M.31 in Andromeda), but it is not at all easy to decide upon the shape of a system when one is living inside it. During the war, H. C. van de Hulst and his colleagues in Holland worked out that for reasons which need not concern us at the moment, the clouds of cold hydrogen in the Galaxy ought to send out radio emission at a wavelength of 21·1 centimetres, and in 1951 Ewen and Purcell, in America, showed that this really does happen. It became possible to plot the positions and the movements of the hydrogen clouds, and a strong indication of spiral structure was found. We now know that the Sun lies not far from the edge of a spiral arm. I will have more to say about this sort of thing when dealing with outer galaxies.

Hydrogen is by no means the only material spread between the stars, and all sorts of unexpected things have been found recently—water vapour, for instance, and organic molecules. It shows that there is no such thing as truly 'empty space', and the significance of these new discoveries is only just starting to be appreciated. So far as the dust is concerned, the grains are extremely small—of the order of a hundred-thousandth of a centimetre in diameter—and it has been found that they are affected by the well-marked magnetic field of the Galaxy, which also is of tremendous importance to theorists.

This may be the moment to introduce the idea of the two so-called 'stellar populations'. It seems that there is a definite difference between the kinds of objects to be found in and near the spiral arms of the Galaxy, and in the region of the nucleus. In the arms, the brightest stars are very hot and of early spectral type (O and B); there is considerable interstellar material. This makes up what we call Population I. In Population II areas, the most luminous stars are old Red Giants which have evolved off the Main Sequence, and the interstellar matter has been more or less used up, so that presumably star formation has to all intents and purposes ceased. This is the situation near the middle of the Galaxy, and also in the globular clusters. Virtually all the objects in the galactic halo seem to be of Population II—though we must be rather wary; the distinction is not always absolutely clear-cut.

Not surprisingly, it has been found that the Galaxy is rotating round its nucleus. Here again there are complications to be taken into account, but we may be sure that the Sun takes about 225 million years to complete one circuit, a period which is sometimes known unofficially as the 'cosmic year'. The velocity is of the order of 150 miles per second. Stars closer to the nucleus have shorter revolution periods; those further out take longer. Objects in the galactic halo have different kinds of paths, and because they do not move in orbits similar to that of the Sun they seem to be very fast-moving—even though they do not really travel with exceptional speed. They are known as high-velocity stars, and are of Population II.

The length of the cosmic year is so great that nobody can

really appreciate it. One cosmic year ago, the Earth was passing through the coal-forest or Carboniferous period; there were no true trees, and the so-called forests were made up of tall plants of the horsetail variety. There were no men, no mammals and only small reptiles; even the great dinosaurs lay in the future—though there were amphibians, and plenty of dragonflies. Two cosmic years ago, and we are back in the Cambrian period, when the continents were completely barren and life was represented only by small, low-type sea creatures. Three cosmic years takes us into the Pre-Cambrian, before life on Earth began. In fact, the whole story of living things on our world is contained in the last three cosmic years. What will happen to us during the next cosmic year remains to be seen, and may well depend upon our own actions in the immediate future!

Meantime, we are at least in a position to look at the various objects in our Galaxy and decide upon how they are arranged. If we could look at the system from outside, we would be able to see the spiral form. We have found, too, that our Galaxy is not unique—and this leads us on to a consideration of the universe itself.

Chapter Seventeen

THE OUTER GALAXIES

HUMAN VANITY HAS suffered a number of hammer-blows during the past few centuries. First it was clearly shown that the Earth is an unimportant planet rather than being the centre of the universe. Next, the Sun was relegated to the status of a dwarf star in our Galaxy. And then, in our own time, we have found that even the Galaxy itself is insignificant in the universe as a whole. As a system of stars, it is probably rather on the large side; but there is nothing at all exceptional about it.

Sir William Herschel had the glimmerings of the truth when he wrote that the starry or resolvable nebulæ might be "no less than whole sidereal systems" which might "outvie our Milky Way in grandeur". He could not be certain, and never really made up his mind. This was not his fault; in his day the distances of the stars had not been measured, and nobody had any reliable idea of the true scale of the universe.

In 1845 a giant telescope was built at Birr Castle, in Ireland, by a most remarkable astronomer—the third Earl of Rosse. Quite unaided, except by workmen from his estate whom he trained as technicians, Lord Rosse set up a 72-inch reflector, which was far larger than anything previously built (Herschel's largest mirror was a mere 49 inches in diameter). It was cumbersome and awkward to use, but it was certainly effective, and it was put to good use.* Lord Rosse turned it toward one of the 'resolvable nebulæ', M.51, which lies in Canes Venatici, not far from the Great Bear. To his amazement, he was confronted with the picture of a whirlpool of light—a true spiral, unlike anything which had been seen before. Other spirals were soon found, and by 1850 the list had grown to fourteen, though for many years only the Birr reflector was able to show them; other telescopes were not powerful enough. Nowadays, vast numbers of spiral systems are known, though let me hasten to

* I have told the whole remarkable story elsewhere: *The Astronomers of Birr Castle* (Mitchell Beazley, 1972). The episode remains unique. Nothing like it has ever happened before in the history of science, and nothing like it can ever happen again.

add that not all the objects once called starry or resolvable nebulæ are spiral in shape.

Naturally, spirals are seen at all sorts of angles. M.31, the Andromeda system, is not favourably placed; we are looking at it almost edge-on, so that the whirlpool effect is largely lost. If we had a face-on view of it, as with M.51, it would be much more spectacular.

By the end of the nineteenth century the difference between the gaseous and the starry nebulous objects was very clear indeed—but with the spirals and their kind, were we looking at features lying inside our Galaxy, or at external systems? On the whole, opinion had swung away from Herschel's guess, and in a famous history of astronomy written in 1902 by Agnes M. Clerke we find the theory described as "a half-forgotten speculation . . . it becomes impossible to resist the conclusion that both nebular and stellar systems are part of a single scheme".

Obviously, the spirals were so remote that they could show no measurable parallax, and the problem would have been very hard to solve without those convenient variable stars which give away their real luminosities simply by their periods of fluctuation. The original clue was provided by the variables in the southern Small Magellanic Cloud. I have already said something about it—see Chapter 10—but I hope you will forgive a certain amount of repetition, because the whole subject is so vitally important.

The two Magellanic Clouds are, alas, too far south to be seen from anywhere in Europe, but they are very familiar indeed to Australians, New Zealanders and South Africans (Plate VII), because they are easily visible with the naked eye. They could easily be mistaken for detached parts of the Milky Way, and telescopes show that they contain all manner of objects, including variable stars and gaseous nebulæ. Around the beginning of the present century, thousands of photographs of them were taken from Arequipa in Peru, where a large telescope had been set up by astronomers based at Harvard in the United States. When Henrietta Leavitt made a close study of the plates, some interesting facts emerged. Over 1750 variables were found, many of which were Cepheids. Miss Leavitt established that the Cepheids in the Small Cloud which looked the brightest had the longest periods; since, for all practical purposes, all the stars in the Cloud could be regarded as lying at the same distance from us, the longer-period stars were genuinely the more luminous. This led on to the famous Cepheid period-luminosity law. It was not long after this that Harlow Shapley studied the short-period variables in the globular clusters, and drew up the first reliable maps of the Galaxy.

Actually, the stars used by Shapley were not classical Cepheids, but the variables which were once called cluster-Cepheids and are now known as RR Lyræ stars. The difference seemed to be unimportant at the time, though it proved to be of tremendous significance later on.

What could be done for the globular clusters could presumably be done also for the resolvable nebulæ, and this would settle the age-old question of whether or not they were external systems. Shapley himself believed that they were not, as he stated in a famous debate with another great astronomer, Heber D. Curtis, in 1920. Yet on this occasion Shapley was wrong. Three years later there came the most far-reaching discovery of all, made by Edwin Hubble with the aid of the 100-inch reflecting telescope at Mount Wilson—then much the most powerful in the world (it was not surpassed until the completion of the Palomar 200-inch, after the end of the last war). Hubble detected Cepheids in the Andromeda Spiral, worked out their distances, and announced that they must be something like 750,000 light-years away. This would put them well outside the Milky Way system; and it followed that the Andromeda Spiral must be a galaxy in its own right.

I doubt whether it is possible to over-emphasize the importance of this discovery. It changed all Man's ideas about the universe; the last illusions of grandeur were shattered 'at a stroke', to use a modern phrase.

Before long it was found that Hubble's original estimate was too low, and that 900,000 light-years would be better. Even so, it appeared that the Andromeda Galaxy was much smaller than ours, and it was generally believed that the Milky Way system must be a sort of super-galaxy; yet there were uneasy doubts. Something might still be badly wrong.

The next step was taken in 1952 by the late Walter Baade,

this time using the Palomar reflector. (It had to be this great telescope; no other instrument had sufficient light-grasp.) The method of attack was highly ingenious. First, Baade made a systematic search for RR Lyræ variables in the Andromeda Spiral, and failed to find any. This was surprising; though RR Lyræ stars are less luminous than Cepheids, they are still very powerful, and at a distance of less than a million light-years they ought to have shown up. Therefore, either they were absent from the Spiral—which did not seem likely—or else the Spiral itself must be so far away that they could not be seen. Finally, Baade found the answer. There are two kinds of Cepheids, Types I and II; a Type I variable is more luminous than a Type II star of the same period. The Cepheids in the Andromeda Spiral had been regarded as belonging to Type II, but in fact they were of Type I, so that all the distance-estimates had to be doubled. The modern estimate for the Andromeda Galaxy is 2,200,000 light-years.

I well remember the meeting of the Royal Astronomical Society at which this result was announced. It caused a sensation, to put it mildly. There had been an error of around one hundred per cent; the universe was twice as large as had been thought—since, of course, the error applied to all the galaxies, not only to M.31. Also, M.31 itself proved to be larger than the Milky Way system rather than smaller.

Fittingly, the two Magellanic Clouds provided extra proof. They contain globular clusters, but up to then there had been a curious anomaly; though similar in form to our own globulars, those in the Clouds had been thought to be only a quarter as brilliant, but if the Clouds themselves were further away there would be no problem. The modern value for their distance is 180,000 light-years. Both systems are genuinely smaller than our Galaxy; the Large Cloud is about 40,000 light-years in diameter and the Small Cloud 20,000, as against 100,000 light-years for the Milky Way system. Both Clouds are more or less irregular in shape; the distance between their centres is 75,000 light-years, and radio studies have shown that they are contained in a common envelope of very rarefied hydrogen.

It has been suggested, though without real conviction, that the Clouds are satellite galaxies of our own system. In any case, they are much the closest of the main external systems, and they can be studied in great detail, so that they are of unique importance to theorists. In fact, this is one reason why really large telescopes are being set up in the southern hemisphere.

The Andromeda Spiral, too, has companion galaxies, one of which (M.32) was listed by Messier; but unlike the Clouds, both these are elliptical in shape, and seem to be predominately Population II, so that their brightest stars are reddish. Hot Main Sequence stars of high luminosity are scarce, though there are plenty of them in the Spiral itself.

The Andromeda Spiral is the most distant object clearly visible with the naked eye. (A famous trick question is: 'How far can you see without optical aid?' The answer is—2·2 million light-years, or roughly 12,000,000,000,000,000,000 miles!) Unfortunately it is not spectacular when seen through a modest telescope, and this also applies to the other galaxies. To study them properly, one has to make use of photographs taken with very powerful instruments, and telescopes of the size owned by most amateur astronomers will not show the spiral forms at all. Look at M.31 by all means, but do not be disappointed to find that it seems only like a smudge of light. Not far from it in the sky is another spiral, M.33 in the constellation of Triangulum (Plate VIII), which is seen more or less face-on, but is much looser and less well-defined as well as being smaller and fainter. Some people claim to have glimpsed it with the naked eye; I have never been able to do so, but it can be identified with binoculars. The central condensation is much less marked than with M.31.

The Milky Way system, the Andromeda and Triangulum Spirals, and the Magellanic Clouds are very close on the distance-scale of the universe, and are contained in what we call the Local Group of galaxies (Fig. 76). There are more than two dozen other systems in the Group, but these are relatively small and loose; for instance, the two dwarf systems seen in Sculptor and Fornax respectively make up a total of only about one per cent of the mass of our Galaxy. However, we must remember that we cannot see all the way through the galactic nucleus, because of the obscuring material in the way—and there may well be other systems beyond permanently hidden from us. There have been suspicions that two objects in Perseus, known as Maffei 1 and Maffei 2 in honour of their discoverer,

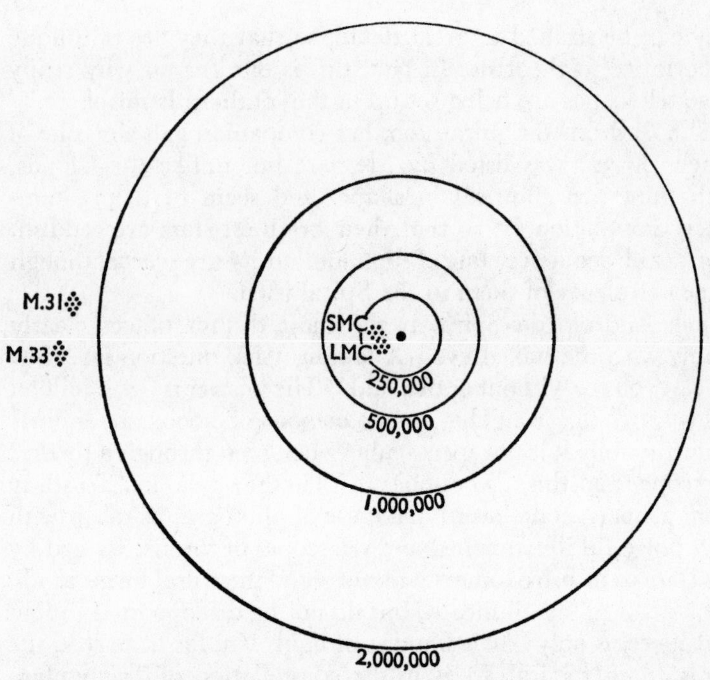

Fig. 76. The Local Group: the two Magellanic Clouds (LMC, SMC) and the Andromeda and Triangulum Spirals (M.31, M.33). Distances are in light-years.

may be massive galaxies in the Local Group; but nobody really knows, because they are too heavily obscured to be properly classified.

Though Cepheids are so luminous, they cannot be seen individually beyond distances of a few million light-years, and so we must look for some other method of distance-measuring. Supergiant stars can be pressed into service. They outshine even the Cepheids, and it seems reasonable to assume that the supergiants in an external galaxy are about equal to those in our own Galaxy—so that we can work out how far away they must be merely by measuring their apparent magnitudes and making the usual corrections. The results are probably fairly reliable, though less so than for the Cepheids, and the method can take us out well beyond the Local Group.

Within the range of the supergiant method we come to a whole group of galaxies called the Virgo Cluster; it contains well over a thousand members, and for most practical purposes these galaxies can be taken to be at the same distance from us. They appear fairly close together in the sky, but this, of course, is misleading; and do not forget that what we call a cluster of galaxies is very different from a star-cluster, either loose or globular.

The Virgo cluster lies at around 40 million light-years from us, and there are systems of all kinds, including spirals. We can calculate the real sizes of these galaxies, and apply the same principles as with the Cepheids and the supergiants, suitably modified. Measure the apparent size of a still more remote galaxy, compare this with what should be its true size—and its distance follows.

Again we have a reduction in accuracy, because even if two galaxies are similar in form they are not necessarily equal in diameter, but the method is better than nothing at all, and it can carry our 'space soundings' out to over two thousand million light-years. Sometimes, of course, extra checks are possible. Supernovæ occur in other galaxies beside our own; if we assume an average maximum luminosity for a supernova, we can estimate its distance.

Generally speaking, the average distance between galaxies seems to be several million light-years, but it is important to bear in mind that except in our own part of the universe we can see only the brighter systems. If the Magellanic Clouds, for instance, lay at 500 million light-years instead of less than 200,000, we would have no hope of seeing them at all, though a major galaxy would still be on view. The tendency toward clustering is so marked that many astronomers believe the true picture to be one of vast arrays of groups, with genuinely isolated galaxies being rare exceptions.

With really remote clusters of galaxies—such as that in Hercules (Plate IX), or Hydra, at over 2,000 million light-years —no structure can be seen; the systems appear only as fuzzy patches on our photographic plates, and it takes an effort of imagination to realize that each is a vast system containing thousands of millions of stars together with clusters, nebulæ and, almost certainly, planetary systems as well.

This is as far as purely optical methods can take us, and to estimate distances which are still greater we have to turn to the spectroscope. As so often happens, we come back to our reliable ally, the Doppler effect. In 1920 V. M. Slipher, at the Flagstaff Observatory in Arizona, examined over four hundred galaxies spectroscopically, and made a curious discovery. Red Shifts were not only the rule, but almost the invariable rule. Nearly all the galaxies were running away from us at high speeds (remember that a Red Shift indicates a velocity of recession). At that time it was still uncertain whether or not the 'starry nebulæ' were external systems, but on all counts the Red Shifts were very much of a problem.

When Hubble had managed to show that the spirals and their kind were true galaxies, he turned his attention to the Red Shifts. Of course, the spectrum of a galaxy is bound to be confused; it does not come from a single body, but is the result of the spectra of all kinds of objects jumbled together, so that the main lines only can be made out. However, the Doppler effect shows up unmistakably, and what Hubble found made the situation even stranger. There was a definite relationship between Red Shift and distance; the more remote galaxies were receding the faster, and the speeds were fantastic, amounting to thousands of miles per second. Only the systems in our Local Group were exempt. It appeared that every cluster of galaxies must be receding from every other cluster; in other words, the universe is expanding.

Let us suppose that the distance/velocity relationship (the 'Hubble Constant') is valid out to as far as we can see. We then have a new method of distance-gauging. Look at the spectrum; measure the Red Shift; work out the velocity of recession—and you have the distance. This, in essence, is how the distances of very far-away galaxies are measured, and the method takes out to systems which are at least 5,000 million light-years off. If the Red Shifts are not pure Doppler effects, then we are obviously in serious trouble; but this a point to which I will return later on. For the moment, let us turn back to the varied forms and characteristics of individual galaxies.

Spirals are of various kinds, from loosely-wound forms to 'tight' Catherine-wheels (Plate X). Hubble originally divided them into three types—Sa, Sb and Sc, as shown in Figure 77. In Sa galaxies, as with M.65 in Leo—at a distance of 29 million light-years—the central condensation is large, with the arms tightly coiled. Sb systems have arms which are more prominent; to this class belongs M.64, nicknamed the 'Black-Eye Galaxy' in Coma Berenices, near the Great Bear. Here the distance is 44 million light-years, but M.64 is larger than most of its companions in the Virgo cluster, and is an easy object in a small telescope—though do not expect to see the spiral arms. Among Sc galaxies, with looser arms, we have the Whirlpool' M.51 and also the Triangulum Spiral, M.33. Today we also recognize galaxies of type SO, which have been described as spiral systems without any true spiral arms. A good example is M.84, again in the Virgo cluster. Then, too, we have the Seyfert galaxies, first noted by Carl Seyfert in 1940, where the nuclei are almost starlike. One of these is M.77, in Cetus, which is over 50 million light-years away and has the distinction of being the senior galaxy of Messier's catalogue. Its total mass has been estimated at 800,000 million times that of the Sun; and like other Seyferts it is a strong source of radio waves.

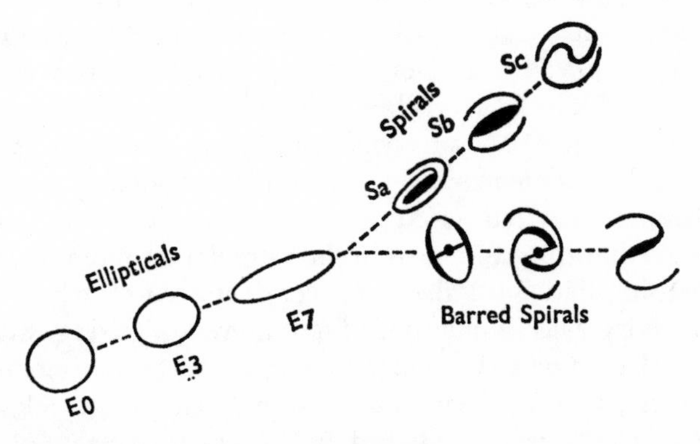

Fig. 77. Hubble's Classification of Galaxies: Ellipticals (Eo to E7), Spirals (Sa to Sc) and Barred Spirals.

In addition there are the extraordinary barred spirals (SBa, SBb, SBc) in which the arms seem to extend not from a true nucleus, but from the ends of a straight bar lying in the plane of the system and passing through its centre. These galaxies are less common than normal spirals in the proportion of about one to three; all the same, there are plenty of them.

Dwarf irregular galaxies are numerous enough; the junior

members of the Local Group belong to this class, and are made up chiefly of Population II. Large irregulars are rare, but there are many systems which are elliptical; here we have sub-grades ranging from E7 (very flattened) down to E0 (spherical). Again Population II predominates.

Probably the most famous of the giant elliptical systems is M.87 (Plate XI), yet again belonging to the Virgo Cluster and lying at 41 million light-years from us. It is of type E0, and from it issues a curious 'jet' which looks very much as though it were formed by material being ejected at a high speed. M.87 is almost the equal of the Seyfert galaxy M.77 in mass, but it has another characteristic too; it sends out about ten thousand times as much energy at radio wavelengths than it might be expected to do—and this brings me on to the whole question of radio galaxies.

We have already noted that many of the radio sources identified in the early days of this new branch of astronomy were extragalactic. One of these was Cygnus A, in the Swan, which was soon tracked down to a faint optical object; improved equipment showed that there were two radio sources lying to either side of the faint visible patch, and it was only in late 1973 that R. J. Peckham, at Manchester, managed to track down much weaker radio signals from the optical object itself. The distance was estimated at 700 million light-years, and it was found that the gas-clouds in the source were moving around at something like 300 miles per second, so that clearly tremendous activity was going on. Even more surprising was the southern Centaurus A, at a much lesser distance (12 million light-years), which when photographed with large telescopes looked very curious indeed, and gave every impression of being compound.

Some explanation had to be found, and for a few years it was thought that we were observing cosmical collisions on a grand scale. If two galaxies met head-on, the individual stars would seldom hit each other, because they are so widely spaced; but the tenuous material between the stars would be colliding all the time, and would, it was believed, account for the radio emission. Around 1960 the idea of colliding galaxies was so deeply-rooted that to question it was almost heretical. Then, as so often happens, new calculations showed that the whole concept would have to be thrown overboard. The amount of energy being sent out is so great that mere collisions of interstellar atoms and molecules will not explain it.

The trouble is that up to now, at least, no really satisfactory alternative has been put forward, but that tremendous disturbances are taking place there can be no doubt at all. For instance, there is the case of M.82 in Ursa Major, which has been described as an exploding galaxy, and is known to have a powerful magnetic field.

In 1961 C. R. Lynds, in America, was studying the larger Sb-type spiral M.81, which is fairly close ('only' 8½ million light-years away!) when he found that a second, weaker radio source, which had been assumed to be associated with this spiral, actually came from its fainter neighbour M.82, which is genuinely irregular—as had been noted with the great 72-inch Rosse reflector as long ago as 1871. Clearly there was something unusual about it, and R. Sandage used the Palomar reflector to show that there were huge, intricate structures of hydrogen gas, moving about at speeds of up to 600 miles per second. It now seems that a colossal outburst took place near the centre of M.82 in the past. From the present movement of the gas, we can fix the date reasonably well; it corresponds to about 1½ million years ago—though since the galaxy is 8½ million light-years from us, the actual date of the explosion goes back a full ten million years. There are not many hot bluish stars in M.82, and it is generally believed that the radio waves are due to what is known as synchrotron emission—that is to say, energy sent out by the acceleration of high-velocity electrons in a strong magnetic field. Incidentally, M.82 is small as galaxies go. Apart from M.32, the elliptical companion of the Andromeda Spiral, it is actually the least massive of all the galaxies in Messier's list.

It would be idle to claim that we yet know a great deal about the evolution of galaxies, and we are not even sure how or why spiral arms form—though it has been shown, at long last, that a spiral rotates with its arms 'trailing', again in the manner of a Catherine-wheel. Elliptical galaxies are mainly of Population II, and so are older than spirals with their hot Population I stars; but it would be very rash to assume that a spiral evolves into an elliptical. There is much that we do not know, and it may even be that in the near future we will have to revise our ideas almost as drastically as astronomers had to do in 1923, when Hubble first showed that the 'spiral nebulæ' are huge systems far beyond our own.

Chapter Eighteen

QUASARS

BEFORE THE 1960s, we were modestly confident that even if we did not know all about the objects in the stellar sky we did at least know all the types of objects which were there. We have been sadly disillusioned. Pulsars came upon the scene before the end of the decade, but even before that there had been the discovery of the remarkable and puzzling 'quasars'.

The story really began with some attempts by astronomers at Palomar to identify a number of small, reasonably intense radio sources which had been listed in the third catalogue drawn up at Cambridge, in England. R. Minkowski, one of the Palomar observers, had been particularly interested in the problem, and had scored a notable success with one of the sources, 3C-295 (that is to say, the 295th object in the Cambridge list). Originally there had been no known object in the position of the radio source, which lies in Boötes, but Minkowski took special photographs with the 200-inch reflector and tracked the emissions down. In the precise position of the radio source there lay a dim galaxy, whose spectral lines were so strongly red-shifted that the distance was obviously very great. In fact 3C-295, at some 5,000 million light-years, turned out to be the most remote object ever identified up to that time.

Encouraged, Minkowski turned his attention to three other sources: 3C 48, 196 and 286. This time he had no luck at all. There were no visible galaxies, but in each case there seemed to be a faint blue star in the approximate position of the radio source. With 3C-48, the sixteenth-magnitude star was accompanied by a wisp of nebulosity. The star seemed to be rather out of the ordinary; definite lines were found in its spectrum, but nobody could identify them.

Early in 1963 the problem was attacked by Maarten Schmidt, also at Palomar. He concentrated upon the source known as 3C-273, because he knew its exact position in the sky. Normally, it is not easy to pinpoint a radio source as accurately as can be done for a star; but it so happens that

3C-273 can be occulted by the Moon, and this had happened three times in the previous year, so that very accurate measurements had been possible.

When the Moon passes in front of a star, and hides or occults it, the star's light seems to snap out like a candle-flame in the wind. The disappearance is virtually instantaneous, because the Moon has no atmosphere, and a star is to all intents and purposes a point source. Occultations have been used to check the position of the Moon, because the stars have been plotted on the celestial sphere with amazing precision—and if you know the moment when the star's light is cut off, you also know where the Moon's edge is at that moment.

With a radio source, the situation is reversed, because this time it is the position of the Moon's limb which is known with the greater accuracy. Also, a radio source is generally of appreciable size, so that it disappears gradually as the Moon sweeps across it. Predictions for 3C-273 enabled C. Hazard and his colleagues at the Parkes radio astronomy observatory, in Australia, to find out just where the radio source lay. In fact, they discovered that there were two sources, separated by about 20 seconds of arc. Using the positions measured at Parkes, Schmidt found that one of them coincided with a star slightly brighter than the thirteenth magnitude, while the more extended source could be identified with a fainter wisp or jet of bluish nebulosity.

The next step was to study the optical spectrum of the brighter source. There was no real difficulty about this—a star of magnitude 13 is bright enough to yield an excellent spectrum —but at first Schmidt was baffled. There were six visible dark lines and three broader emission regions, superimposed on a background continuous spectrum which was very bright in the blue region. But what about the identification of the lines? They did not seem to fit in with anything at all.

Then, one evening, Schmidt decided to see whether he could identify the lines if he applied a correction for a Doppler shift to either the blue or the red end of the spectrum. Immediately he began work he had a tremendous shock. All six lines could be identified (four with hydrogen, one with oxygen and one with magnesium) if he were prepared to admit a 16 per cent shift of the red, which would mean a tremendous velocity of recession—and, therefore, a vast distance. Once the clue had been found, the spectrum of 3C-48 was examined in the same way, with the same result. Again there was a huge Red Shift. Again the main lines could be confidently identified as being due to hydrogen.

If these results were reliable—and of this there seemed little doubt—then 3C-273, 3C-48 and other similar objects could not be stars, and they could not be members of our Galaxy. Since they were so unexpected, nobody had a name ready for them; at first they were called QSOs (a conventional abbreviation for Quasi-Stellar Objects), but by now they have become known as quasars. This is more comprehensive, and also more appropriate, since it has been found that not all quasars are powerful emitters of radio waves.

As soon as these startling facts became widely known, the great Quasar Hunt began. Dozens were found; by now the list runs into hundreds, and we cannot question the reality of the optical identifications. We know what quasars look like, and we also know that some of them are variable, particularly in the radio range (3C-273, for instance, fluctuated to the extent of 30 per cent in two years). What we have not so far been able to discover is an answer to the all-important question: What are they?

Assuming that the Red Shifts in their spectra are Doppler effects, as with galaxies, all quasars are not only very remote but are also super-luminous. A powerful quasar may outshine a complete galaxy of the Milky Way type—and, remember, a galaxy may contain more than a hundred thousand million stars. Yet a quasar looks more or less stellar, so that compared with a galaxy it is extremely small. Its diameter cannot be more than a few light-years, so that we have to work out how so much energy can come from so diminutive an object. At first glance the situation makes no sense at all, and neither do the velocities of recession. A quasar thousands of millions of light-years from us, well beyond the 5,000 million light-years of the radio galaxy 3C-295, measured by Minkowski, may be racing away at a velocity not so very far short of that of light.

Let me make one point clear immediately. There are some astronomers who doubt whether the Red Shifts are pure Doppler effects, in which case the quasars would be neither so remote nor so powerful as has been generally thought; but even so, they must still lie well beyond our Galaxy, and they must be highly luminous. In passing, they show no measurable proper motions, and they could not be expected to do so. In a way it is misleading to say that they were 'discovered' in 1962–3; they had been known long before, and some of them were recorded on photographs taken in the last century, but they had always been taken for ordinary stars (as can be seen from Plate XII). There was no reason to suspect otherwise—until radio astronomy entered the field. A few nights ago I turned the 15-inch reflector in my own observatory toward 3C-293. It looked remarkably innocent and stellar, and had I not been able to identify it, from its position on my charts, I would have had not the slightest chance of realizing that there could be anything unusual about it.

We seem to be faced with an energy-source which is new to us, and about which we know regrettably little. So first let us try to decide whether the colossal luminosities and distances indicated by the spectroscopic evidence are genuine or not. If they are not, then at least the quasars will become slightly less incredible.

It is no use trying to deny the Red Shifts themselves; the spectral lines are perfectly clear. We know that Doppler effects can produce shifts of this kind. There can also be red-shift effects due to gravitation—and such shifts have been measured for some very small, very dense, relatively nearby White Dwarfs such as the companion of Sirius. It would be convenient to say simply that the Red Shifts in quasars are gravitational, but Schmidt and his colleague J. Greenstein have shown that the very strong gravitational fields needed would prevent the emission of some of the characteristic spectral lines that we can observe. At best, then, gravitational red-shifting can play only a minor rôle. We are forced back to the idea that velocities of recession must be responsible; we know of no other mechanism which would do.

And yet could there be a loophole? It has been suggested that quasars might have been shot out from various galaxies, including our own, by some Titanic explosion, so that they have been hurled outward in all directions; in other words they are genuinely receding as quickly as their Red Shifts indicate, but no overall expansion of the universe is involved. There is no known quasar with a spectral shift to the violet or short-wave end, and this means that our hypothetical explosion must have occurred quite close to our Galaxy. But just imagine the violence of the outburst needed! The débris would surely have been spread out through space more or less evenly; any local condensations would have been ripped apart at an early stage—and as the material expanded, the formation of large masses would become less and less likely. It is inconceivable that the end product would be localized, very massive objects. All in all, this 'local hypothesis' simply does not work, and by now there are very few astronomers who have any faith in it.

Quasars, then, are extremely remote and powerful, but we cannot be sure that the laws regulating their Red Shifts are the same as with the galaxies. Some contrary evidence has come from the work of the American astronomer Halton Arp, who has found that there are some significant alignments of some quasars with strings of galaxies. If the two classes of objects are associated, then they should have the same Red Shifts; but they do not—those of the quasars are much the greater. This leads to the awkward speculation that we might have similar distances but different Red Shifts; and if we accept the shifts of the galaxies as being pure Doppler effects, then some other factor must be involved for the quasars. It is fair to say that some of Arp's alignments have been questioned, but in 1973 it was found that with two quasars very close together in the sky there was a similar difference in red-shifting. The chances that the twin quasars are at different distances from us are very slight, and if any more similar pairs are found we may have to do some radical re-thinking. At present the whole question remains open.

For the moment I propose to accept the majority view that the distances of the quasars are exactly as predicted on the pure-Doppler hypothesis, but definite doubts remain. Moreover, it is true to say that all our measurements of distances in the further reaches of the universe depend upon the Red Shifts; and if these prove to be misleading—well, we may not be back in Square One, but we will certainly have to return to Square Two.

Let me repeat that on the cosmological picture a quasar may be far more luminous than any galaxy, but it has a much smaller diameter—only a few light-years at most, as against 100,000 light-years for the Milky Way system. The tongue of matter coming from 3C-273, which is the brightest and presumably the nearest of all the quasars, may be 150,000 light-years long, but even this is not very much on the cosmical scale; it would not, for instance, reach from the Sun out as far as the Magellanic Clouds. But if a quasar were even remotely comparable in size with a galaxy, it would appear as a blob, not a starlike point. This at once disposes of the idea that a quasar can be simply a remote galaxy with more than its fair share of highly luminous stars.

Another suggestion, made in the early days, was that the energy-source might be a series of supernovæ; one such outburst would trigger off another, and so on. This seems to have nothing to recommend it. There is no known way in which such a thing could happen, and no supernova stays at peak brightness for more than a few months; quasars are longer-lived than that, and must persist for a million years at least. Yet another idea was due to Sir Fred Hoyle, in 1964, who proposed that a quasar might be caused by the sudden creation of new matter out of nothingness. On the old steady-state theory of the universe (to which I will turn in the next chapter), matter appears spontaneously in the form of single atoms spread over a wide area; if by some unknown process an immense mass could come into being in the same area at the same time, a quasar could result. However, this too has been discarded. There are many objections to it, and in any case the whole steady-state theory now seems to be defunct.

Next there is the anti-matter theory, due to the great Swedish scientist Hannes Alfvén. Basically, it is supposed that there are two kinds of matter; the kind which makes up our own part of the universe (koinomatter, to use Alfvén's term) and a diametrically opposite kind (antimatter). When an atom of koinomatter meets an atom of anti-matter, the two annihilate each other and produce a flash of energy. Spectroscopically, the two types would be indistinguishable, so that we would have no means of knowing whether a remote galaxy were made up of anti-matter or not. Alfvén has developed the theory in detail, and suggests that between koinomatter and anti-matter regions there are 'boundary layers' where a certain amount of mutual annihilation is always going on. Now, if a koinomatter galaxy met an anti-matter galaxy, the mutual destruction would be tremendous, and would certainly yield enough energy to produce a quasar.

This is all very well, but the whole picture is so speculative that most astronomers view it with suspicion. We have absolutely no evidence that the basic principle is valid, and there is no obvious way of finding out. Therefore, it is best to regard it only as an ingenious and not impossible hypothesis. (Science fiction writers have made great play of it. Suppose a koinomatter boy fell in love with an anti-matter girl? Fond embraces would be a little hazardous!)

To show how fluid our ideas are, let me give a very brief account of two theories, each of which has some backing from eminent theorists. The first is that a quasar is a very young galaxy, in the process of formation; in fact a quasar and a Seyfert galaxy represent two stages of evolution of the same class of object. Intense internal activity might—it is suggested—provide enough energy for a quasar, though there would have to be some unknown additional factors involved. Alternatively, it has been proposed that a quasar is a galaxy which is coming to the end of its life, and that it has a rotating black hole in its centre; the black hole is gobbling up the remaining stars round its rim, so that the stellar matter which manages to escape is given extra energy sufficient to produce a quasar-like appearance. On a modification of this rather depressing picture, a quasar contains many small black holes, which are merging together with the emission of vast quantities of radiation.

Very young or very old? We do not know. Recently there has been some slight evidence of galaxy-like features associated with at least some quasars; and if this can be confirmed it will indicate that a quasar really is of 'galaxy' nature, though we are still faced with the problem of deciding why it is so powerful. In America, J. B. Oke and J. Gunn, using the 200-inch reflector, have carried out a long study of the strange object known as BL Lacertæ, which they find to be a quasar embedded in a normal large galaxy. This discovery is so recent that its full significance may not yet be apparent, but it certainly is of major importance.

Whichever way we turn, quasars seem to be strange enough to make any astronomer adopt the attitude of Lewis Carroll's White Queen, who made a habit of believing at least six impossible things before breakfast every day. It is ironical that pulsars, which were discovered six years later and which caused just as much of a sensation, were satisfactorily explained quite quickly; nobody now doubts that they are neutron stars. But the quasars remain as puzzling as ever. All we can really say is that they are definitely outside our own Galaxy, probably lying at distances of thousands of millions of light-years; that they are relatively small compared with ordinary galaxies; and that their immense power is due to some process which could well involve gravitational energy; and that they may be contained inside normal galaxies.

Up to now we have had less than fifteen years of quasar research, and perhaps it is too optimistic to hope for a rapid solution. Energetic studies are going on all the time, and we can only hope that some of the answers will be found before long. I have the feeling that if we can only solve the great quasar problem, we will have made a major breakthrough in our understanding of the universe itself; but as yet we may not even know what the main problem really is.

Chapter Nineteen

THE UNIVERSE

MEN HAVE ALWAYS wondered how the universe began. This is natural enough; in a way it is the most fundamental of all problems, and if we could solve it we would be well on the way to real wisdom. Unfortunately this is easier said than done, and during the past few decades arguments have raged not only between scientists and theologians but also between eminent scientists whose ideas are as opposite to each other as the proverbial chalk and cheese. Cosmology, the study of the history of the universe, can be highly controversial, and the arguments have sometimes tended to become heated.

Before going any further, let me deal briefly with the theological side of the problem. The main argument originally presented was delightfully simple. In the seventeenth century Archbishop Ussher of Armagh stated blandly that the world came into existence at 10 a.m. on 26 October, 4004 B.C. He came to this conclusion by adding up the ages of the patriarchs and making various other calculations of the same kind. The figure was at once accepted by the Church, and remained official for many years. Alas, it was clear to any scientist that the world is a great deal older than that; the evidence of fossils, in particular, was conclusive. I am not sure when the Church conceded that Ussher's date must be wrong, but it went very much against the clerical grain to concede that the Earth is thousands of millions of years old. I may add that some people question it even now; there is still an Evolution Protest Movement, just as there is still a Flat Earth Society!

Ideas of this kind are ludicrous today, but they were not ludicrous three hundred years ago, and it would be quite unfair to laugh at Ussher. What he did, of course, was to claim that the universe was divinely created at a set moment in time; and who can prove otherwise? The fact that his date of 4004 B.C. was much too recent is neither here nor there.

I refrained from calling this chapter "The Creation of the Universe", which would have been the obvious title, because I am something of a rebel in this respect, and scientifically I do not believe that anyone has ever discussed the actual origin of the universe. The various big-bang, steady-state, cyclic and other theories, to be discussed below, relate to the evolution of the universe, which is a very different thing; and if I may be forgiven for over-simplification, the situation can be summed up in a very few words. The one inescapable fact is that the atoms and molecules making up you, me, this book, Aunt Emily, the Sun, the Moon, the stars, the quasars and everything else *exist*. They must have come from somewhere, and there seem to be only two possibilities:

1. The material making up the universe came into existence at one definite moment, which may be called the beginning of time: or

2. The material has always existed, so that the universe had no beginning, and will presumably have no end.

Unfortunately, both these alternatives lead us into appalling difficulties. If (1) is correct, and 'time' began, then what happened earlier? But if we follow (2), we are faced with picturing a period of time which had no beginning, which is equally impossible—and would, I feel, be impossible even to the White Queen. In fact, we have not the faintest idea about the original creation. All we can do is to assume that matter existed, and work out an evolutionary sequence, perhaps beginning with a uniform diffuse gas and ending up with the universe as it is today.

Actually, the trouble may well be that we are asking ourselves an unfair question. The universe is four-dimensional, and we are three-dimensional creatures. Whether we will ever make enough progress to come to any definite conclusions remains to be seen, but there is no immediate prospect of it.

So far as everyday language is concerned, we are in a similar quandary with regard to the size of the universe. If space is finite, then what lies beyond it? But if not, then we have to imagine something which goes on for ever—and, as before, we are a dimension short.

There is one aspect of current research which can give a slight feeling of unease. In our studies of the universe, a tremendous amount depends upon the recession of the galaxies, and here we rely almost entirely upon the Red Shift. Practically all astronomers agree that the galaxies beyond our Local Group are racing away from us, and that the shifts are Doppler effects; but just suppose that there is an error in interpretation? If the galaxies are not receding, then most modern cosmological theories fail at once. Before the discovery of the quasars (in particular, the recent twin quasar pair) anyone who doubted the expansion of the universe was regarded in much the same light as a member of the Evolution Protest Movement. Even today the number of dissentients is very small, but it is not nil, and only when the quasar riddle has been cleared up will we be absolutely confident.

For the moment, however, I will assume—as most people do—that the Red Shifts of both galaxies and quasars indicate universal expansion, and I will also assume that the Hubble Constant remains valid out as far as we can see. (To recapitulate: the Hubble Constant links distance with recessional velocity, so that the most remote galaxies are moving away with the greatest speeds.) Working backwards, so to speak, we can find that in the past the galaxies must have been much closer together than they are now, whether or not they then existed in the forms we know; and if the Hubble Constant has not altered in the meantime, it seems that expansion cannot have been going on for much more than 10,000 million years. This gives us something of a basis, and we do know, of course, that the ages of the Earth and the Moon are between four and a half and five thousand millions years, which fits in quite well.

One of the first men to draw up a modern-type theory of the early days of the universe was a Belgian priest, the Abbé Lemaître. Lemaître began with a very dense 'primæval atom' comprising all the material in the universe. The density of this primæval atom would have to be tremendous; there could be no proper elements as we know them, and certainly no individual stars or galaxies. Then, between 20,000 million and 60,000 million years ago, the primæval atom exploded, sending its material outward in all directions. Expansion, the direct result of this outburst, went on for thousands of millions of years, until the whole universe had a diameter of something like 1,000 million light-years. At this stage things began to settle down, and clusters of galaxies began to form from the primæval material.

We know that gravitation tends to draw material together, and one might imagine that as soon as the force of the explosion had spent itself the matter in the universe would move once more toward a common centre. Lemaître supposed that this did not happen for the excellent reason that there is another force, cosmical repulsion, which acts in the contrary fashion to gravity over very great distances, though over small distances —such as those within our Solar System, our Galaxy or even the Local Group—it is negligible. If so, the 'settling-down' universe would be in a state of balance, cosmical repulsion just counteracting gravity and preventing either a general expansion or a general contraction. Then, about 9,000 million years ago, some new disturbance tipped the scales in favour of cosmical repulsion; expansion began, and has continued ever since, because a larger universe means that cosmical repulsion will grow stronger while the opposing force becomes weaker.

Lemaître's theory was publicized and improved by Sir Arthur Eddington, one of the greatest of all pioneers of cosmology. A further modification was made by George Gamow, Russian by birth but American by adoption. This time there was no need for cosmical repulsion; the present expansion was attributed solely to the force of the initial 'big bang'. Gamow worked out the early temperatures very precisely. Five minutes after the start of expansion the temperature was 1,000 million degrees, and after a day (that is to say, twenty-four hours) had dropped to 40 million degrees, falling steadily and reaching a stable value millions of years later. Gamow also believed that all the chemical elements so familiar to us now were formed within half an hour after the universe began. There was no long-drawn-out state of balance, as in Lemaître's theory.

And yet there was still the vexed question of "What happened before the beginning?" and in 1948 a group of astronomers at Cambridge came out with a daring new theory which abolished the 'beginning' entirely. Instead, they produced an hypothesis which involved the concept of continuous creation. The original papers were due to H. Bondi and T. Gold; later modifications were introduced by Hoyle and others.

Let us go back to the expansion of the universe. Light travels at 186,000 miles per second; the most remote galaxies known are moving away at velocities not far short of this—and the further away they are, the faster they go. Eventually, then, we will reach a distance when galaxies are receding at the full velocity of light. We cannot see such systems at all, no matter how strong our equipment; their light will never reach us, and they will have passed over the boundary of the observable universe.

On the 'big bang' theory, the sky in the very remote future will look very different from that which we know. Since the galaxies are moving away from us at velocities which increase as the distances from us increase, there must come a time when all of them will have passed beyond the limit of the observable universe, in which case the sky will be empty of galaxies except those of the Local Group. Of course, the time-scale is very long. As Eddington once said, we need be in no hurry to study the galaxies before they vanish from our ken!

The Cambridge group rejected this scheme entirely. According to their steady-state or continuous creation theory, the universe has always been in much the same state as it is now, and will never change. As old galaxies pass over the observational boundary, new galaxies will appear in their stead. It follows that material is being created out of nothingness all the time, and that the average density of matter in any part of the universe remains constant.

It is not supposed that a new galaxy can appear ready-made in an instant. Matter is created in the form of hydrogen atoms, and the rate of creation is very slow, though over the whole observable universe it amounts, in tons, to a figure 1 followed by 32 zeros per second. When hydrogen is created, the cycle begins; with majestic slowness the material collects, and galaxies form. On the steady-state theory, then, an intelligent being living in the far future will see the same number of galaxies as we do today—but they will not be the same galaxies.

Such was the steady-state picture. It must, I think, have caused more controversy than any theory since Darwinian evolution, but there seemed to be no possible direct test, because matter created at such a rate would be harder to track down than a single new grain of sand on Bognor beach. But there was one indirect test which, it was thought, might be good enough to show whether or not the universe were in a steady state. This was by studies of the distribution of very far-away galaxies.

Now, remember that when we look out into space, we are also looking backward in time. Light from the Moon takes 1¼ seconds to reach us, so that we always see the Moon as it used to be 1¼ seconds ago; the Sun appears in its form of 8½ minutes ago, while the corresponding figures are over 4 years for Proxima Centauri, 27 years for Vega, about 900 years for Rigel and so on. The Andromeda Spiral is seen as it used to be 2,200,000 years ago; but what about a really remote system, such as the famous radio galaxy 3C-295? Here we are looking back 5,000 million years, to a time before the Earth existed. And 5,000 million years is a very appreciable fraction of the time which seems to have elapsed since the start of the expansion of the universe.

On the original steady-state theory, the average amount of material in any given volume of space has always been approximately the same, so that if we could board a time-machine and go back into the past we would always find the overall aspect similar to that which we know. The galaxies would have the same sort of distribution. Not so on the evolutionary or big-bang theory, because thousands of millions of years ago the galaxies were crowded together, giving a completely different picture.

We cannot build a time-machine, but—and this is the crux of the whole problem—we can look back in time, simply by studying the remote galaxies. If the distribution is the same in those regions as it is nearer home, then the steady-state theory is presumably right. If the distribution is different, then the universe is not in a steady state, and the theory is wrong.

When the Cambridge astronomers first put forward their revolutionary ideas, such a test could not be applied, because no telescope in the world was powerful enough to reach out to such distances; the Palomar 200-inch was only just coming into use. Moreover, radio astronomy was in its infancy, and quasars were as unknown as they were unsuspected. The first hint of decisive evidence came in 1955, when the team led by Professor (now Sir) Martin Ryle published some counts of faint radio sources. Radio waves can be studied over greater distances than waves of visible light, and of course they move at the same velocity.

The early results showed that the distribution of remote

radio sources was markedly different from anything to be expected on the steady-state hypothesis. Naturally there were fierce arguments, and it is true that the preliminary counts were later shown to be highly suspect; but later work pointed in the same direction, and the steady-state theorists had to fight an uphill battle. Various modifications were introduced. At one stage, I remember, it was suggested that we might be living in an expanding part of a steady-state universe (!) and, with more reason, doubts were cast upon the nature of the radio sources upon which the counts were based.

Quasars, of course, came into the controversy from 1963 onward. In 1966 Dr. Dennis Sciama, who was then a steady-state supporter, began to study the numbers of quasars with different Red Shifts to see whether or not they fitted into the expected pattern. He and his colleagues found that there were many too many quasars with large Red Shifts to fit in with the model. True, there were (and are) doubts as to whether the quasars are reliable guides in this sort of research, since it is just possible that their Red Shifts are not connected with the general expansion of the universe; but it soon became clear that the evidence was impressive even without the quasars.

Next came a fortuitous discovery which dealt the steady-state idea a further blow. In 1965 two radio astronomers in America, A. Penzias and R. Wilson, were observing the sky with special radio equipment, and found that there was considerable 'noise' from outer space, apparently coming in from all directions. At Princeton University, Robert Dicke and his team had been busy making a new radio telescope designed to search for precisely this kind of effect. Their reason ran along the following lines: The original expansion of the universe could have been preceded by a collapse from a still earlier dilute phase (widely-spread gas, presumably), and during this collapse the temperature of the universe would have been raised. After the big bang occurred, the heat decreased. Dicke calculated that approximately one minute after the explosion the overall temperature would have been a thousand million degrees, and that by now it would have dropped to about three degrees above absolute zero, i.e. −270°C. As soon as he heard of the work by Penzias and Wilson, Dicke suggested that the radiation they had found was nothing more nor less than the background radiation which was left over from the big bang.

Of course, I realize that I have over-simplified the problem very seriously; but it does look as though the background radio noise has all the characteristics which would be expected on Dicke's theory, and most astronomers incline to the view that the theory itself is correct. If so, then steady-state is not. Plausible and elegant though it may be, it fails to fit the facts, and we must return to the picture of a universe which is evolving.

On this basis, the next question to be asked is: Will the expansion continue indefinitely? If the answer is 'yes', then presumably there has been only one big bang, and the galaxies (or, more correctly, the groups of galaxies) will continue to recede from each other until they have moved out of all contact. We can visualize a universe in the very far future, with each group of galaxies being left on its own; and eventually the universe will die—with nothing left but dead, lightless stars. Yet this, too, has been challenged. It may well be that the universe is cyclic, so that expansion is slowed down prior to a new stage of shrinking; eventually all the material comes together again, so that there is another big bang, and the story is repeated. There is no reason why this should not have happened many times in the past.

Again we must rely upon radio astronomy to give us the first evidence, positive or negative. If the Hubble Constant is unchanged out to the very edge of the observable universe, then there can never be another contraction; but it is possible that the very distant galaxies are receding less quickly than they would do if the Hubble figure applied in its accepted form. If this is true, then there will come a moment when the furthest galaxies will start to draw together again, and in perhaps 60,000 million years' time the universe will be reborn in another incredibly hot 'fireball'. Here, of course, we have an alternative to cosmic death, because the expansion-and-contraction process could be repeated time and time again. Rather irreverently, I always think of this as the Concertina Theory.

From the theoretical viewpoint, much depends upon the amount of material in the universe: is it enough to draw the galaxies back before they escape from each other's pull? If we consider only the stars and other visible material the answer seems to be 'no', but there may be a great deal of material about which we know nothing, and once more the black hole concept comes into the discussion. If tremendous quantities of material are contained inside black holes, then there is no reason why the total mass of the universe should not be more than enough to stop the expansion before it is too late. Remember that although we cannot see black holes, their gravitational pulls are just as strong as they would be if we could actually observe them.

At the moment, we must concede that our ideas are still in a state of flux. The choice today seems to be between either indefinite expansion and final death, or a cyclic universe which could continue for ever. Philosophically, the second picture is obviously the more attractive, but we must await further confirmation—or denial.

This, I fear, is as far as we can go as yet. The fundamentals still elude us; but we have learned more than would have seemed even remotely possible a few years ago, and we may hope that one day we will find the answers.

Chapter Twenty

TRAVEL TO THE STARS?

FLIGHT TO THE Moon seemed impossible only a few decades ago, and the very idea of a probe to, say, Mars or Jupiter seemed to be wildly futuristic. Reputable scientists were saying that nothing of the sort could ever be managed. So far as interstellar travel is concerned, the same is true today, and I admit to being one of the pessimists. Yet I must stress that when I cast doubts upon the feasibility of reaching the stars, I am not suggesting that it will always remain impossible. All I am saying is that it is impossible by our present techniques, which is a very different thing.

The trouble, basically, is one of sheer distance. The Apollo astronauts completed their lunar trips in less than a fortnight; the Mars probes took only a few months on their way; Pioneer 10 needed eighteen months to by-pass Jupiter. When we develop nuclear-powered rockets, these times can be cut, and I have no doubt that we will eventually be able to explore the whole of the Solar System, at least by using unmanned space-ships.

But let us go back to our scale model. If the Earth-Sun distance is represented by one inch, then the distance to Proxima Centauri is over four miles; and neither Proxima nor the other members of the Alpha Centauri system seem to be likely centres of planetary systems. Of course one cannot be sure; but if we are searching for a planet similar to the Earth, it is only reasonable to start at a star similar to the Sun, and the two nearest candidates (Tau Ceti and Epsilon Eridani) are both about eleven light-years from us.

Unfortunately we cannot see planets of other stars, because of the distance problem. However, if we adopt any one of the current theories about the origin of the Solar System, according to which the planets grew up by accretion out of a cloud of material associated with the youthful Sun, it is surely reasonable to suppose that other stars of the same kind have produced planet-families. It is equally logical to think that an Earth-type planet will support Earth-type life, and most modern astronomers consider that life is likely to be widespread throughout the universe.

The only possible way of finding out, by our 1974 techniques, is to establish radio contact with some other civilization, and this is by no means easy! The first serious attempt was made in 1960, by Dr. Frank Drake and his colleagues at Green Bank in West Virginia, using the 85-foot radio telescope there. The project was known officially as Ozma, though I understand that the participants usually referred to it as Operation Little Green Men.

After careful consideration, Drake selected a wavelength of 21·1 centimetres for his main search. As we have noted, radiation at this wavelength is sent out by interstellar hydrogen, and radio astronomers, wherever they may be, will be interested in it. There could be no hope of picking up intelligible messages; we could hardly expect other beings to understand English—but a rhythmical pattern of signals, showing that the transmissions were non-natural, would be a tremendous step forward. To narrow the search still further, Drake concentrated upon the two nearest stars which are reasonably like the Sun: Tau Ceti and Epsilon Eridani.

There was no response, and I doubt whether anything else could have been expected. The project was abandoned after a few months, though sporadic attempts have been made more recently in the Soviet Union. There have also been some sensational reports, such as that of 13 April, 1965, when one British newspaper column on the front page was headed: "Astounding Claim by Russian Astronomers. Life in Space: Is it Superman?" Actually the signals came from a quasar, CTA-102; and a later report, in 1973, was equally disappointing. Journalists, like flying saucer enthusiasts, do tend to jump to conclusions.

The whole problem of extra-terrestrial life is very complicated, and a discussion of it would be out of place here. Suffice to say, then, that although we do not know much about the fundamental nature of life, we do know a great deal about the make-up of living matter; and so far as we can tell, the essential atom is that of carbon. There are indications that life, wherever it may be, must be carbon-based, in which case it will need an equable temperature, air to breathe and water to drink. Once we depart from this concept, we enter the science-fiction realm of Bug-Eyed Monsters; and though I would be the last to deny that such creatures *may* exist, all the current evidence is against them. If there are B.E.M.s, then most of our modern science is wrong, and this is something which I for one am most reluctant to believe.

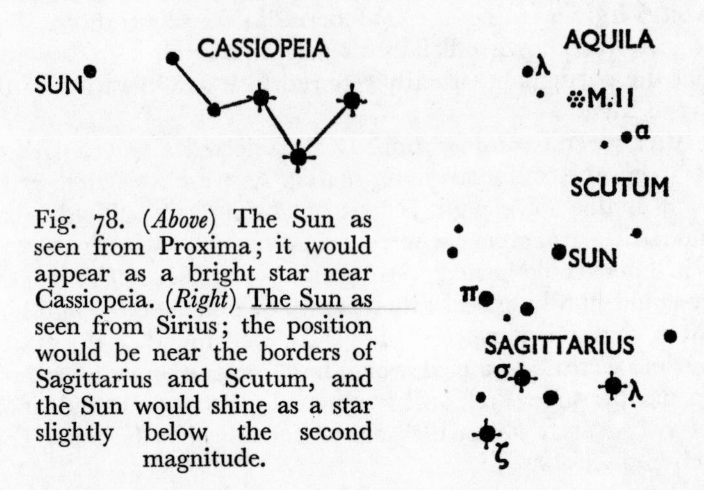

Fig. 78. (*Above*) The Sun as seen from Proxima; it would appear as a bright star near Cassiopeia. (*Right*) The Sun as seen from Sirius; the position would be near the borders of Sagittarius and Scutum, and the Sun would shine as a star slightly below the second magnitude.

Therefore, we may justifiably assume that other civilizations will have at least some points in common with ours; and for all we know, they may use radio and optical telescopes of the same kind. The Sun would not show any signs of eminence when seen from afar. From Proxima, for instance, it would be a fairly bright star near the W of Cassiopeia; from Sirius it would be of the second magnitude, and would lie on the borders of Sagittarius and Scutum (Fig. 78). (Because both Proxima and Sirius are so near, the constellation patterns made up of much more remote stars would be much the same.) The Earth, and even Jupiter, would be much too faint and close to the Sun to be detected with, say, a telescope comparable with our Palomar 200-inch reflector.

Though we cannot see planets of other stars, there is another method of attack. A massive planet orbiting a relatively modest and nearby star might cause perturbations marked enough to be detected. There are several reported cases, the most convincing being that of Barnard's Star, which is only six light-years away, and is a dim Red Dwarf with a mere 1/2,300 of the Sun's luminosity. It has the largest proper motion of any known star—10·3 seconds of arc per year, so that in 170 years it crosses a distance equal to the apparent diameter of the full moon. Tiny irregularities in this proper motion have led Peter van de Kamp and his colleagues in America to conclude that it is attended by two planets, each about the same mass as Jupiter. This seems reasonably conclusive, but let us remember that a planet no more massive than the Earth would be pitifully inadequate to cause measurable disturbances in the proper motion of any star.

It would be rather pointless to launch out upon an interstellar voyage without the assurance of finding a suitable planet on arrival, and of this there is no prospect as yet. But in any case, the difficulties of journeying to the stars are immense in comparison with a local trip inside the Solar System. Present-day rockets would be useless; so would the nuclear space-ships which should be built during the 1980s. Centuries would be needed to reach even the nearest star.

No doubt the space-ships of, say, A.D. 2100 will be capable of greater speeds, but even at the velocity of light it would still take four years to reach Proxima and eleven to get to Tau Ceti or Epsilon Eridani. Moreover, there are theoretical objections to travel at such a velocity. According to Einstein's theory of relativity, which has so far survived every test, there are strange effects when the speed of light is approached; one's mass increases and one's time slows down. Both these remarkable phenomena have been confirmed to some extent. For instance, cosmic rays, which are high-speed particles coming from outer space, smash into the Earth's upper air, creating other particles which we call mu-mesons; these mu-mesons decay so quickly that they ought to have no time to reach the ground—and yet they do so, because they are moving so fast that their time, relative to ours, is slowed down.

This brings us on to what is called the Twin Paradox. Consider identical twins, one of whom elects to go on an interstellar journey at 99 per cent of the velocity of light. The traveller departs, and is away for a few years, during which interval he visits a neighbouring stellar system. When he comes home, he finds that his twin has become an old man, because the traveller's time-scale has been slowed down compared with that of Earth. If taken further, this means that the traveller could be away for what he would imagine to be a decade or so, and then return to be greeted by his great-great-grandchildren.

This time-dilation effect has caused a tremendous amount of argument, and by no means all authorities accept it at face value. There may be a fundamental fallacy somewhere. In any case, the effect does not become appreciable except at very high velocities; so far as travel to the Moon and planets is concerned, we can ignore it completely. No vehicle which we can think of building as yet could move fast enough to make its planners take the time-dilation phenomenon into account.

It is, I suppose, conceivable that we will eventually be able to construct space-ships which will work up to an appreciable fraction of the velocity of light; but the critical value of 186,000 miles per second (or, if you want to be precise, 186,282·3959 miles per second or 670 million m.p.h.) can never be reached by a material object. This, to me, indicates that we can never achieve interstellar travel by using rockets.

Various ideas have been proposed, all of them rather peculiar. The 'space-ark' is a science fiction favourite; the original pioneers, and several succeeding generations, die during the voyage, and only their descendants survive to reach their target (by which time, I suspect, the purpose of the entire journey would have been forgotten). On a second plan, the crew are deep-frozen and put into a state of suspended animation until the time comes for them to wake up and prepare to land on, say, Epsilon Eridani B. I can only say that I would not personally care to attempt anything of this sort, and I cannot believe that the human mind and body would tolerate it.

No: if we are to travel between the stars, it must surely be by some non-material technique. It may sound fantastic to talk about teleportation or even thought-travel, but no more so than television would have seemed to Julius Cæsar or King Canute. I am quite ready to accept that something of the sort will come along one day, but as yet we have nothing to guide us, so that speculation is both endless and pointless.

Of course, it may be that interstellar civilizations will contact us before we are able to contact them. Mankind is still young, and there is plenty of time ahead of us unless we are stupid enough to turn our planet into a radioactive waste. As yet we are isolated from all the other beings who must exist in the universe, but this isolation may not continue indefinitely. Time will tell.

Meanwhile, we have at least sent one probe into deep space. In December 1973 the automatic space-craft Pioneer 10 passed within 90,000 miles of Jupiter, and by now it has well and truly begun a never-ending journey. It has escaped from the Solar System, and it will never return. It carries a plaque, just in case some alien race finds it and wants to find out from whence it came; the chances of its ever being picked up are millions to one against—but one never knows. And it is a measure of our progress that an Earth-made vehicle is at this moment travelling outward in the space between the stars.

Appendix I

THE CONSTELLATIONS

Latin Name	*English name*	*Leading stars*	*Remarks*
Andromeda	Andromeda	Alpheratz	
Antlia	The Airpump		Very low in Britain
Apis	The Bee		Invisible in Britain
Aquarius	The Water-bearer		Zodiacal
Aquila	The Eagle	Altair	
Ara	The Altar		Invisible in Britain
Aries	The Ram	Hamal	Zodiacal
Auriga	The Charioteer	Capella	
Boötes	The Herdsman	Arcturus	
Cælum	The Sculptor's Tools		Low in Britain
Camelopardalis	The Giraffe		
Cancer	The Crab		Zodiacal
Canes Venatici	The Hunting Dogs	Cor Caroli	
Canis Major	The Great Dog	Sirius	
Canis Minor	The Little Dog	Procyon	
Capricornus	The Sea-Goat		Zodiacal
Carina	The Keel	Canopus	Invisible in Britain
Cassiopeia	Cassiopeia		
Centaurus	The Centaur	Alpha Centauri, Agena	Invisible in Britain
Cepheus	Cepheus		
Cetus	The Whale	Diphda, Mira	
Chamæleon	The Chameleon		Invisible in Britain
Circinus	The Compasses		Invisible in Britain
Columba	The Dove		Low in Britain
Coma Berenices	Berenice's Hair		
Corona Australis	The Southern Crown		Invisible in Britain
Corona Borealis	The Northern Crown	Alphekka	
Corvus	The Crow		
Crater	The Cup		
Crux Australis	The Southern Cross	Acrux	Invisible in Britain
Cygnus	The Swan	Deneb	
Delphinus	The Dolphin		
Dorado	The Swordfish		Invisible in Britain
Draco	The Dragon		
Equuleus	The Little Horse		
Eridanus	The River	Achernar	Largely invisible in Britain
Fornax	The Furnace		Very low in Britain
Gemini	The Twins	Pollux, Castor	Zodiacal
Grus	The Crane	Alnair	Invisible in Britain
Hercules	Hercules		
Horologium	The Clock		Invisible in Britain
Hydra	The Watersnake	Alphard	
Hydrus	The Little Snake		Invisible in Britain
Indus	The Indian		Invisible in Britain
Lacerta	The Lizard		
Leo	The Lion	Regulus	Zodiacal
Leo Minor	The Little Lion		
Lepus	The Hare		

Latin name	*English name*	*Leading stars*	*Remarks*
Libra	The Scales		Zodiacal
Lupus	The Wolf		Invisible in Britain
Lynx	The Lynx		
Lyra	The Lyre	Vega	
Mensa	The Table		Invisible in Britain
Microscopium	The Microscope		Very low in Britain
Monoceros	The Unicorn		
Musca Australis	The Southern Fly		Invisible in Britain
Norma	The Rule		Invisible in Britain
Octans	The Octant		Invisible in Britain
Ophiuchus	The Serpent-bearer	Rasalhague	
Orion	Orion	Rigel, Betelgeux	
Pavo	The Peacock		Invisible in Britain
Pegasus	The Flying Horse		
Perseus	Perseus	Mirphak, Algol	
Phœnix	The Phœnix	Ankaa	Invisible in Britain
Pisces	The Fishes		Zodiacal
Piscis Austrinus	The Southern Fish	Fomalhaut	
Puppis	The Poop		Partly invisible in Britain
Pyxis Nautica	The Mariner's Compass		Low in Britain
Reticulum	The Net		Invisible in Britain
Sagitta	The Arrow		
Sagittarius	The Archer	Kaus Australis	Zodiacal
Scorpio	The Scorpion	Antares	Zodiacal
Sculptor	The Sculptor		Low in Britain
Scutum	The Shield		
Serpens	The Serpent		
Sextans	The Sextant		
Taurus	The Bull	Aldebaran	Zodiacal
Telescopium	The Telescope		Invisible in Britain
Triangulum	The Triangle		
Triangulum Australe	The Southern Triangle		Invisible in Britain
Tucana	The Toucan		Invisible in Britain
Ursa Major	The Great Bear		
Ursa Minor	The Little Bear	Polaris	
Vela	The Sails		Invisible in Britain
Virgo	The Virgin	Spica	Zodiacal
Volans	The Flying Fish		Invisible in Britain
Vulpecula	The Fox		

(Carina, Vela and Puppis are the parts of the old Argo Navis, the Ship Argo, which was dismembered because of its vast size. Of the remaining constellations, Hydra has the greatest area and Crux Australis the least.

A few constellations have alternative names [Scorpio may be Scorpius, while Ophiuchus was once known as Serpentarius], and many of the now-accepted names are abbreviations of the old ones; thus Vulpercula was once Vulpecula et Anser, the Fox and Goose, while Cælum was originally Cæla Sculptoris. Ophiuchus is not ranked as a member of the Zodiac, but it does nevertheless cross the Zodiacal band between Scorpio and Sagittarius.)

Appendix II

THE BRIGHTEST STARS

Star		*R.A.* *h*	*m*	*s*	*Decl.* °	′	″	*Magnitude*	*Spectrum*	*Remarks*
Sirius	Alpha Canis Majoris	06	43	50	−16	40	25	−1·43	Ao	White Dwarf companion
Canopus	Alpha Carinæ	06	23	17	−52	40	44	−0·73	Fo	Highly luminous
—	Alpha Centauri	14	37	33	−60	42	46	−0·27	Go +K5	Fine binary
Arcturus	Alpha Boötis	14	14	18	+19	20	16	−0·06	Ko	Lovely orange star
Vega	Alpha Lyræ	18	35	55	+38	45	17	0·04	Ao	Distinctly bluish
Capella	Alpha Aurigæ	05	14	28	+45	58	10	0·09	Go	Very close binary
Rigel	Beta Orionis	05	13	06	−08	14	06	0·15	B8	Highly luminous
Procyon	Alpha Canis Minoris	07	37	44	+05	18	11	0·37	F5	White Dwarf companion
Achernar	Alpha Eridani	01	36	36	−57	23	20	0·58	B5	
Betelgeux	Alpha Orionis	05	53	33	+07	24	10	var.	Mo	Range about 0·1–1·0
Agena	Beta Centauri	14	01	41	−60	13	45	0·66	B1	
Altair	Alpha Aquilæ	19	49	19	+08	47	16	0·80	A5	
Aldebaran	Alpha Tauri	04	34	12	+16	27	01	0·85	K5	Lovely orange star
Acrux	Alpha Crucis	12	24	55	−62	55	59	0·87	B1 +B1	Fine binary
Antares	Alpha Scorpii	16	27	34	−26	22	01	0·98	Mo	Companion, type A3
Spica	Alpha Virginis	13	23	37	−11	00	19	1·00	B2	
Fomalhaut	Alpha Piscis Austrini	22	56	00	−29	46	54	1·16	A3	
Pollux	Beta Geminorum	07	43	29	+28	06	00	1·16	Ko	Orange star
Deneb	Alpha Cygni	20	40	24	+45	10	21	1·26	A2	Highly luminous
—	Beta Crucis	12	45	57	−59	31	30	1·31	B1	
Regulus	Alpha Leonis	10	06	46	+12	06	52	1·36	B8	
Castor	Alpha Geminorum	07	32	41	+31	57	19	1·58	Ao +A2	Fine binary

These are the stars usually classed as being of the 'first magnitude', though the limit is unofficial; many lists omit Castor. There are also minor differences in the magnitude values in published lists; thus Aldebaran has been measured at 0·78, Rigel at 0·08 and so on. These discrepancies are much too slight to be detectable with the naked eye.

[R.A.=Right Ascension; Decl.=Declination.]

Appendix III

THE NEAREST STARS

Star	*R.A.*		*Decl.*		*Parallax*	*Distance*	*Spectrum*	*Magnitude:*	
	h	*m*	°	′	″	*lt.-yrs.*		*Apparent*	*Absolute*
Proxima	14	26	−62	28	0·762	4·2	M5	+10·7	+15·1
Alpha Centauri A	14	36	−60	38	0·751	4·3	G0	0·0	+4·4
Alpha Centauri B	14	36	−60	38	0·751	4·3	K5	+1·4	+5·8
Barnard's Star	17	55	+04	24	0·545	6·0	M5	+9·5	+13·2
Wolf 359	10	54·1	+07	20	0·402	8·1	M8	+13·5	+16·5
Lalande 21185	11	00·7	+36	18	0·398	8·2	M2	+7·5	+10·5
Sirius A	06	43·9	−16	40	0·375	8·7	A0	−1·4	+1·4
Sirius B	06	43·9	−16	40	0·375	8·7	A5 (Wh. Dwarf)	+8·5	+11·4
UV Ceti A	01	36·4	−18	13	0·369	9·0	M6	+12·5	+15·3
UV Ceti B	01	36·4	−18	13	0·369	9·0	M6	+13·0	+15·8
Ross 154	18	46·7	−23	53	0·351	9·3	M6	+10·6	+13·3
Ross 248	23	39·5	+43	56	0·316	10·3	M6	+12·2	+14·7
Epsilon Eridani	03	30·6	−09	38	0·303	10·8	K2	+3·7	+6·1

These are the only stars within eleven light-years of us. No giants are included, and only Sirius A and Alpha Centauri A are more luminous than the Sun. Epsilon Eridani is a K-type Main Sequence star, while the rest—apart from Sirius B, of course—are feeble Red Dwarfs.

Appendix IV

SOME INTERESTING DOUBLE STARS

Star	*R.A.* h	m	*Decl.* °	′	*Mags.*	*P.A.*	*Separation* ″	*Remarks*
Beta Tucanæ	00	29·3	−63	14	4·5, 4·5	170	27·1	Both components are close doubles
Eta Cassiopeiæ	00	46·1	+57	33	3·6, 7·5	293	10·1	Very slow binary
Gamma Arietis	01	50·8	+19	03	4·8, 4·8	359	8·2	Very easy fixed pair: 'Twins'
Alpha Piscium	01	59·4	+02	31	4·3, 5·2	297	2·1	Slow binary
Gamma Andromedæ	02	00·8	+42	06	2·3, 5·1	063	10·0	Orange, bluish. B is double
Omicron Ceti	02	16·8	+03	12	var, 10·0	131	0·8	Mira
Iota Cassiopeiæ	02	24·9	+67	11	4·7, 7·0 7·1	240, 116	2·3 8·2	Triple star. Little change in distance or P.A.
Theta Eridani	02	56·3	−40	30	3·4, 4·4	088	8·5	Fine pair
Beta Orionis	05	13·1	−08	14	0·1, 7·0	206	9·2	Rigel. Fairly easy
Theta Orionis	05	32·8	−05	25	Multiple; the Trapezium in the Orion Nebula, M.42			
Beta Monocerotis	06	26·4	−07	00	4·6, 4·7	132	7·4	Brighter component again double
Alpha Geminorum	07	32·6	+31	57	2·0, 2·9	variable		Castor. Binary: 350 years
Gamma Leonis	10	17·2	+20	06	2·6, 3·8	122	4·3	Binary, 407 years. Fine pair
Alpha Crucis	12	24·9	−62	56	1·6, 2·1	114	4·7	Acrux. Very easy pair
Gamma Virginis	12	39·1	−01	11	3·6, 3·6	variable		Binary, 180 years. Sep. now 5″
Alpha Canum Ven.	12	53·7	+38	35	2·9, 5·4	228	19·7	Cor Caroli
Zeta Ursæ Majoris	13	31·9	+55	11	2·4, 3·9	150	14·5	Mizar. Naked-eye pair with Alcor
Alpha Centauri	14	37·5	−60	43	0·0, 1·7	variable		Superb binary: 80 years
Epsilon Boötis	14	42·8	+27	17	2·7, 5·1	338	2·9	Yellowish, bluish
Beta Scorpii	16	02·5	−19	40	2·9, 5·1	023	13·7	Fine pair
Nu Scorpii	16	09·1	−19	20	4·3, 6·5	337	41·4	Both components are again double
Alpha Scorpii	16	27·4	−26	22	1·1, 6·5	274	2·9	Antares. Companion greenish
Zeta Herculis	16	39·4	+31	42	3·1, 5·6	variable		Binary, 34 years; widest 1″.6
Alpha Herculis	17	12·4	+14	27	var, 5·4	109	4·6	Companion greenish
Nu Draconis	17	31·2	+55	13	4·9, 5·0	312	62·0	Very wide and easy
Epsilon Lyræ	18	42·7	+39	37	4·7, 4·5	172	207·8	Quadruple; each component double
Zeta Lyræ	18	43·0	+37	32	4·3, 5·9	150	43·7	Very easy
Theta Serpentis	18	53·8	+04	08	4·5, 4·5	103	22·6	Very easy. Perfect twins
Beta Cygni	19	28·7	+27	52	3·2, 5·4	055	34·6	Yellow, blue-green. Superb
Gamma Delphini	20	44·3	+15	57	4·5, 5·5	269	10·4	Yellowish, greenish

Appendix V

SOME INTERESTING VARIABLE STARS

Star	R.A. h	m	Decl. °	′	Range mag.	Period days	Spectrum	Type
R. Andromedæ	00	21·4	+38	18	5·9–14·9	409	S	Mira type
Gamma Cassiopeiæ	00	53·7	+60	27	1·6– 3·2	—	B (pec.)	Irregular
Zeta Phœnicis	01	06·3	−55	31	3·6– 4·1	1·67	B7	Eclipsing
R Arietis	02	13·3	+24	50	7·5–13·7	187	M	Mira type
Omicron Ceti	02	16·8	−03	12	1·7–10·1	331	M	Mira
Rho Persei	03	02·0	+38	39	3·3– 4·2	±40	M	Semi-regular
Beta Persei	03	04·9	+40	46	2·2– 3·5	2·87	B8	Algol. Eclipsing
Lambda Tauri	03	57·9	+12	21	3·3– 4·2	3·95	B + A	Eclipsing
R Leporis	04	57·3	−14	53	5·9–10·5	433	N	Mira type
Epsilon Aurigæ	04	58·4	+43	45	3·3– 4·2	9898·5	F	Eclipsing
Zeta Aurigæ	04	59·0	+41	00	4·9– 5·5	972·1	K + B7	Eclipsing
W Orionis	05	02·8	+01	07	5·9– 7·7	±212	N	Semi-regular
Beta Doradûs	05	33·2	−62	31	4·5– 5·7	9·8	G	Cepheid
Alpha Orionis	05	53·3	+07	24	0·1– 1·0	—	M	Semi-regular
U Orionis	05	52·9	+20	10	5·3–12·6	372	M	Mira type
Eta Geminorum	06	11·9	+22	31	3·1– 3·9	±233	M	Semi-regular
Zeta Geminorum	07	01·0	+20	39	3·7– 4·3	10·2	G	Cepheid
L^2 Puppis	07	12·0	−44	33	2·6– 6·0	141	M	Semi-regular
R Leonis	09	44·9	+11	40	5·4–10·5	313	M	Mira type
U Hydræ	10	35·1	−13	07	4·8– 5·8	—	N	Irregular
Eta Carinæ	10	43·1	−59	25	—0·8– 7·9	—	Peculiar	Irregular
U Carinæ	10	55·8	−59	28	6·4– 8·4	38·8	G	Cepheid
Z Ursæ Majoris	11	53·9	+58	09	6·6– 9·1	198	M	Semi-regular
R Hydræ	13	27·0	−23	01	4·0–10·0	386	M	Mira type
Theta Apodis	14	00·4	−76	33	6·4– 8·6	119	M	Semi-regular
R Centauri	14	12·9	−59	41	5·4–11·8	547	M	Mira type
Delta Libræ	14	58·3	−08	19	4·8– 6·1	2·3	A	Eclipsing
R Coronæ Bor.	15	46·5	+28	19	5·8–15	—	G (pec.)	Irregular
R Serpentis	15	48·4	+15	17	5·7–14·4	357	M	Mira type
T Coronæ Bor	15	57·4	+26	04	2·0–10·8	—	Q + M	Recurrent nova
Alpha Herculis	17	12·4	+14	27	3·0– 4·0	—	M	Semi-regular

R Scuti	18	44·8	−05	46	5·7– 8·6	±144	G to K	RV Tauri type
Beta Lyræ	18	48·2	+33	18	3·4– 4·1	12·9	B	Eclipsing
Kappa Pavonis	18	51·8	−67	18	4·0– 5·5	9·1	G	Cepheid
R Lyræ	18	53·8	+43	53	4·0– 5·0	±47	M	Semi-regular
R Cygni	19	35·5	+50	05	6·5–14·2	426	S	Mira type
Chi Cygni	19	48·6	+32	47	3·3–14·2	407	S	Mira type
Eta Aquilæ	19	49·9	+00	53	3·7– 4·7	7·2	G	Cepheid
P Cygni	20	15·9	+37	53	3 – 6	—	B (pec.)	Nova, 1600
U Cygni	20	18·1	+47	44	6·7–11·4	465	N	Mira type
U Delphini	20	43·2	+17	54	5·6– 7·5	—	M	Irregular
W Cygni	21	34·1	+45	09	5·6– 7·6	±130	M	Semi-regular
Mu Cephei	21	42·0	+58	33	3·6– 5·1	—	M	Irregular
Delta Cephei	22	27·3	+58	10	3·6– 4·3	5·4	G	Cepheid
Beta Pegasi	23	01·3	+27	49	2·3– 2·9	±35	M	Semi-regular
R Pegasi	23	04·1	+10	16	7·1–13·8	377	M	Mira type
Rho Cassiopeiæ	23	55·9	+51	07	4·1– 6·2	—	F to K	Type unknown

Appendix VI

RECENT NAKED-EYE NOVÆ

Year	*Constellation*	*R.A.*		*Decl.*		*Max.*	*Remarks*
		h	*m*	°	′	*mag.*	
1901	Perseus	03	26	+43	24	0·0	GK Persei
1903	Gemini	06	41	+30	00	5·1	
1910	Lacerta	22	33	−52	22	5·0	
1912	Gemini	06	52	+32	12	3·3	
1918	Aquila	18	46	+00	32	0·7	Fast nova
1920	Cygnus	19	57	+53	29	1·8	
1925	Pictor	06	35	−62	36	1·1	RR Pictoris
1934	Hercules	18	07	+45	51	1·3	DQ Herculis. Slow nova
1936	Lacerta	22	14	+55	23	1·9	CP Lacertæ. Fast nova
1936	Sagittarius	18	05	−34	21	4·5	
1936	Aquila	19	24	+07	30	5·0	
1942	Puppis	08	10	−35	13	0·4	Very fast nova
1946	Corona	15	57	+26	04	3·1	T Coronæ; recurrent nova
1960	Hercules	18	55	+13	10	3·0	
1963	Hercules	18	13	+41	50	3·9	
1967	Delphinus	20	40	+18	59	3·7	HR Delphini. Very slow
1968	Vulpecula	19	46	+27	03	4·9	LV Vulpeculæ. Fast nova
1970	Serpens	18	28	+02	40	4·4	

Previous novæ visible with the naked eye have been the stars of 1006 (Lupus; supernova of about magnitude −3), 1054 (Taurus; supernova, about −4; produced the Crab Nebula), 1572 (Tycho's supernova, magnitude −4), 1604 (Kepler's) supernova, magnitude −3), 1670 (Vulpecula, magnitude 3), 1848 (Ophiuchus, 5·5), 1876 (Q Cygni, 3), 1891 (Auriga, 4·5) and 1898 (Sagittarius, 4·7). T Coronæ blazed up to the second magnitude in 1866, and P Cygni rose to the third magnitude in 1600, though it cannot be regarded as a true nova and is still visible with the naked eye.

Appendix VII

SOME CONSPICUOUS CLUSTERS AND NEBULÆ

Object	*R.A.* *h*	*m*	*Decl.* °	′	*Type*	*Remarks*
47 Tucanæ	00	21·9	−72	22	Globular	Magnificent naked-eye globular cluster
M.31 Andromedæ	00	40·0	+41	00	Spiral galaxy	Just visible to the naked eye
NGC 362 Tucanæ	01	00·7	−71	06	Globular	Just visible to the naked eye
H.VII 42 Cassiopeiæ	01	06·0	+58	03	Open cluster	Fairly condensed. Near Phi Cassiopeiæ
M.33 Trianguli	01	31·0	+30	24	Spiral galaxy	Faint, but visible in binoculars
H.VI 31 Cassiopeiæ	01	42·5	+61	00	Open cluster	Fine cluster, visible in finder
H.VI 33 Persei	02	17·2	+56	55 }	Open clusters	Sword-Handle; visible with the naked eye.
H.VI 34 Persei	02	20·4	+56	53 }		Superb with a wide-field eyepiece
M.34 Persei	02	38·8	+42	34	Open cluster	Large open cluster, just naked-eye
M.77 Ceti	02	40·1	−00	14	Seyfert galaxy	Mag. 8·9, and not easy with small telescopes
M.45 Tauri	03	44	+24		Open cluster	Pleiades. Finest of all clusters
M.1 Tauri	05	31·5	+31	59	Supernova remnant	Crab Nebula. Mag. 8·4
M.42 Orionis	05	32·9	−05	25	Great Nebula	
NGC 2070 Doradûs	05	39·1	−69	09	Nebula	Looped Nebula, in Large Magellanic Cloud (round 30 Doradûs). Naked eye
M.35 Geminorum	06	05·8	+24	21	Open cluster	Visible with the naked eye. Fine cluster
M.41 Canis Majoris	06	44·9	−20	41	Open cluster	Just visible with the naked eye
M.44 Cancri	08	37·4	+20	00	Open cluster	Præsepe
M.67 Cancri	08	47·8	+12	00	Open cluster	Just visible with naked eye
Kappa Crucis	12	50·7	−60	05	Open cluster	The Jewel Box; stars of various colours
M.64 Comæ	12	54·3	+21	57	Spiral galaxy	Black-Eye Galaxy. Mag. 6·6
Omega Centauri	13	23·7	−47	03	Globular	Finest of all globulars; very bright
M.3 Canum Venaticorum	13	39·9	+28	38	Globular	Mag. 6·4; bright condensed cluster
M.5 Serpentis	15	16·0	+02	16	Globular	Mag. 6·2; bright condensed globular
M.80 Scorpii	16	14·1	−22	52	Globular	Mag. 7·7. Between Antares and Beta Scorpii
M.13 Herculis	16	39·9	+36	33	Globular	Just visible with the naked eye
M.92 Herculis	17	15·6	+43	12	Globular	Not much inferior to M.13

continued overleaf

Object	*R.A.*		*Decl.*		*Type*	*Remarks*
	h	*m*	°	′		
M.6 Scorpii	17	36·8	−32	11	Open cluster	Naked-eye cluster; splendid sight
M.7 Scorpii	17	50·6	−34	48	Open cluster	Very bright; combined mag. 4
M.8 Sagittarii	18	00·1	−24	23	Nebula	Lagoon Nebula; naked-eye object
M.16 Serpentis	18	16·0	−13	48	Nebula	Easy in binoculars
M.17 Sagittarii	18	17·9	−16	12	Nebula	Omega Nebula; mag. 7. Easy object
M.11 Scuti	18	48·4	−06	20	Open cluster	'Wild Duck'; fan-shaped cluster
M.57 Lyræ	18	51·7	+32	58	Planetary	Ring Nebula, between Beta and Gamma
M.27 Vulperculæ	19	57·4	+22	35	Planetary	Dumb-bell Nebula. Mag. 7·6
H.IV 1 Aquarii	21	01·4	−11	34	Planetary	'Saturn Nebula'; bright, bluish
M.15 Pegasi	21	27·6	+11	57	Globular	Very bright, condensed cluster
M.39 Cygni	21	30·4	+48	13	Open cluster	Large cluster; naked-eye object

Appendix VIII

EQUIPMENT FOR THE AMATEUR OBSERVER

I HAVE STRESSED that so far as amateur astronomers are concerned, the opportunities for scientific research are somewhat limited once we move beyond the Solar System. However, there are some really important lines to be followed up. If you want to search for novæ, then you need powerful, mounted binoculars—but remember that before you have any hope oı success, you must spend literally years in learning the star patterns; not only those visible with the naked eye, but also those which are too faint to be seen without optical aid. Nobody can hope to cover the whole sky in this way, but almost all novæ appear close to the Milky Way zone, and on the whole it is best to concentrate upon selected areas.

For measuring double star separations and position angles, a powerful telescope, with clock drive and micrometer, is a necessity. Variable stars, however, are much more convenient, and I have given some observing hints earlier in this book. Binoculars will suffice for studies of many of the stars in the Appendix V list; for instance, W Cygni, W Orionis and R Scuti can always be seen in binoculars, and so can many of the Mira stars when near maximum—U Orionis and R Leonis, for example.

The larger the telescope, the more variables will come within range. I would recommend something such as a 6-inch reflector, if the cost is to be kept below £100. I would not personally pay much for any reflector below 6 inches aperture, or any refractor below 3 inches; with an outlay of, say, £20 to £30 I would prefer good binoculars. A 7×50 pair is suitable (that is to say, a magnification of 7 with each object-glass 50 millimetres in diameter). Binoculars with a magnification of more than about 12 are awkward to use when hand-held, because their fields of view are so small, but this presents no real problem, because anyone who is reasonably good at carpentry can make a rough mounting.

So my advice may be summed up in a few words. If you want to do some serious observation, presumably of variable stars, then obtain an adequate telescope if you can, but be ready for the need to spend £60 or so; otherwise, use binoculars. If your wish is merely to 'look around', then a smaller telescope will show some lovely sights, though again I would say that binoculars are much to be preferred.

Appendix IX

BIBLIOGRAPHY

AITKEN R. G. *The Binary Stars*. Dover, 1963. A reprint of the classic book which first appeared in 1935.

ALFVÉN, H. *Worlds-Antiworlds*. Freeman, 1966. An account of the fascinating, albeit controversial, anti-matter theory.

GLASBY, J. S. *Variable Stars*. Constable, 1968. A comprehensive account of all kinds of variable stars.

GLASBY, J. S. *The Dwarf Novæ*. Constable, 1970. A more detailed treatment of the U Geminorum or SS Cygni stars.

HEY, J. S. *The Evolution of Radio Astronomy*. Elek, 1973. The story of radio astronomy—by one of its principal founders.

JAKI, S. L. *The Milky Way*. David and Charles, 1971. A description of our Galaxy in all its aspects.

JOHN, L. (Editor). *Cosmology Now*. B.B.C., 1973. A series of articles, based on broadcasts. Contributors include Sir Bernard Lovell, Dennis Sciama and W. H. McCrea.

LOVELL, SIR B. *The Story of Jodrell Bank*. Oxford, 1968.

—— *Out of the Zenith*. Oxford, 1973. Two books which tell the fascinating story of how the world's most famous radio astronomy observatory was established.

MOORE, PATRICK. *The Amateur Astronomer*. Lutterworth; 8th edition, 1974. Intended for the amateur who is anxious to carry out real observation. Includes star maps, lists, etc.

MOORE, PATRICK. *The Southern Stars*. Timmins (Cape Town): Rigby (Sydney, Australia). The observational guide for southern-hemisphere amateurs.

NORTON, A. P. *Star Atlas*. Gall and Inglis, 1973. The classic atlas, covering all the sky; a fully revised text.

SHAPLEY, H. *Galaxies*. Harvard, 1973. An overall survey of the outer systems, written by the man who first measured the size of our own Galaxy.

WHITNEY, C. A. *The Discovery of our Galaxy*. A. A. Knopf, 1971. Mainly historical, and very readable.

INDEX